CHRONICLES
FROM
PAST PLAGUE YEARS

Also by Thomas Fensch ...

Steinbeck and Covici: The Story of a Friendship
Conversations with John Steinbeck
The FBI Files on Steinbeck
Essential Elements of Steinbeck
Steinbeck's bitter fruit:
   from the Grapes of Wrath to Occupy Wall Street
The Man Who Was Dr. Seuss;
   The Life and Work of Theodor Geisel
Of Sneetches and Whos and the Good Dr. Seuss:
   Essays on the Life and Work of Theodor Geisel
The Man Who Was Walter Mitty:
   The Life and Work of James Thurber
Conversations with James Thurber
The Man Who Changed His Skin:
   The Life and Work of John Howard Griffin
Behind Islands in the Stream:
   Hemingway, Cuba, the FBI and the crook factory
Oskar Schindler and his List:
   The Man, the Book, the Film, the Holocaust and Its Survivors
Foreshadowing Trump:
   Trump Characters, Ethics, Morality and Fascism in Classic Literature
Timeless (Pen) Names: The Life and Work of Charles Lutwidge Dodgson, Samuel Langhorne Clemens, Eric Blair and Theodor Geisel
The Kennedy-Khrushchev Letters
Orwell in America
The Books That Haunt Us
War Diaries from Inside Hitler's Headquarters

... and others ...

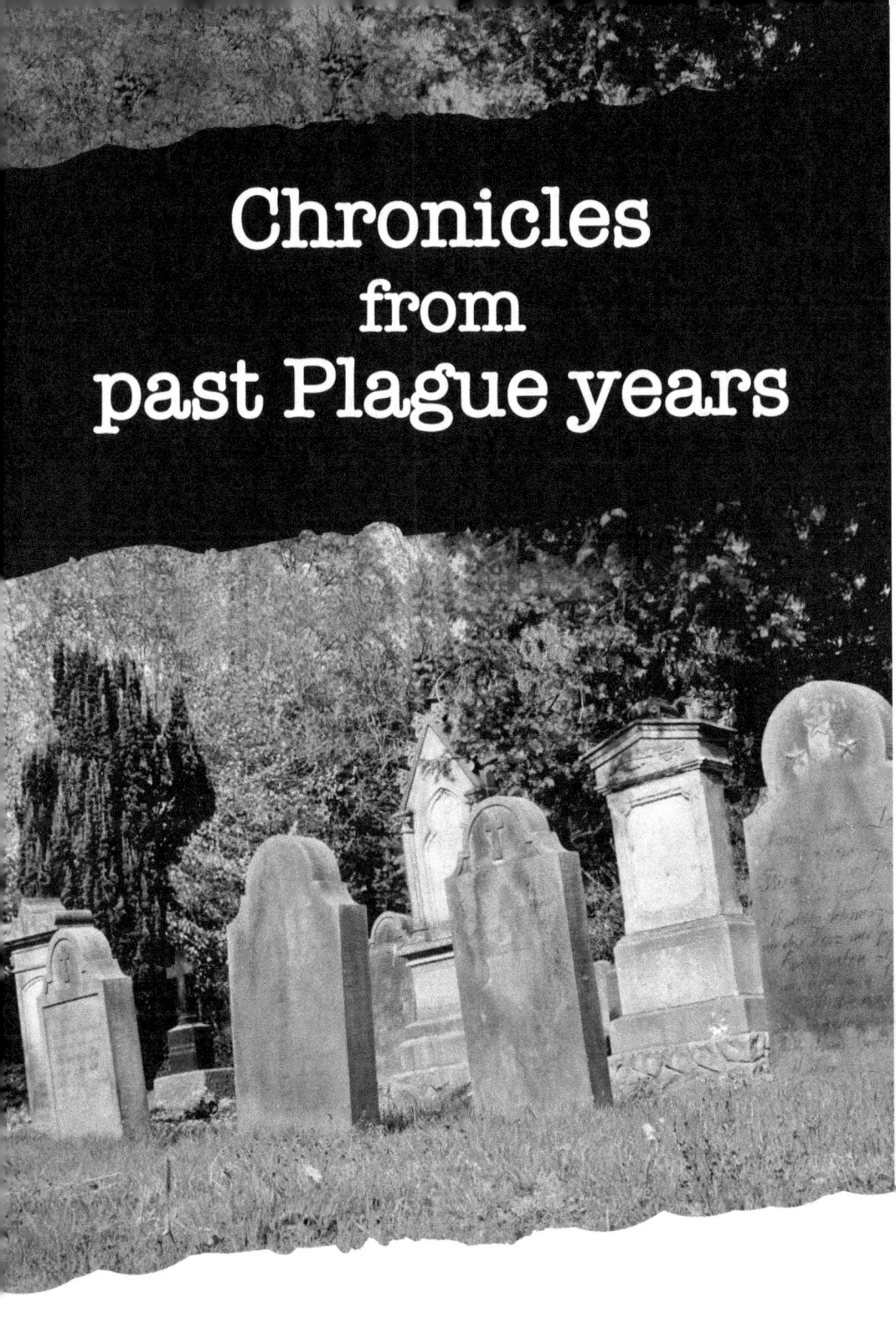

Copyright 2021 Thomas Fensch

All rights reserved. No part of this book may be reproduced or utilized in any form or by any means, electronic or mechanical, including photocopying or recording, or by any other information storage or retrieval system, without written permission from the publisher.

New Century Books
8821 Rockdale Rd.
N. Chesterfield, Va., 23236-2150
newcentbks@gmail.com

ISBN 978-1-7379998-5-0 Hardcover
ISBN 978-1-7379998-1-2 Paperback
ISBN 978-1-7379998-6-7 E-book

Book design by Jill Ronsley, suneditwrite.com
Tombstone cover photograph courtesy Pixabay

Contents

Prologue   i

1. Bubonic Plague, London, 1665-1666
A Journal of the Plague Year, Daniel Defoe, 1722   1

2. The Spanish Flu, 1918-1919
world-wide: no mass vaccines, no antibiotics … no hope   286

3. Bubonic Plague in Algeria
The Plague, Albert Camus, 1947   308

4. AIDS in America
And the Band Played On, Randy Shilts, 1987   317

5. Ebola in America
The Hot Zone, Richard Preston, 1994   346

6. Covid-19
The Plague Year, Lawrence Wright, 2021   358

Epilogue   385
References   386
Bibliography   388
About the author …   390

Prologue

In mid-September 2021, fatalities in the United States from Covid-19 surpassed those of the 1918–1919 Flu pandemic.

Perhaps thinking of the years of the Confederacy and the post-Confederacy South, William Faulkner once wrote, in *Requiem for a Nun*, "The past is never dead. It's not even past."

The same can be said for plagues—they are not dead. They are not even past.

This book begins with the Bubonic Plague that hit London in 1665–1666, that Daniel Defoe described in *A Journal of the Plague Year*—written so many years after the actual event. Whether his *Plague Year* is historical fiction or historical non-fiction or perhaps nascent journalism, is still in debate, but it has left an indelible picture of those times.

The Spanish Flu of 1918–1919 has been described by a variety of writers, including Katherine Anne Porter and Thomas Wolfe.

More importantly—and recently—AIDS, Ebola and Covid-19 have been written, as journalists would say, by those who "covered the stories": AIDS by Randy Shilts; Ebola by Richard Preston and Covid-19 by Lawrence Wright.

There are graphic and horrific stories here about illnesses and deaths during plagues in the past; and in the case of Covid-19—the plague that is with us still.

1

Bubonic Plague, London, 1665-1666

A JOURNAL OF THE PLAGUE YEAR, DANIEL DEFOE, 1722

DANIEL DEFOE, BIRTH DATE UNCERTAIN—ranging from 1659 to 1662, and most probably 1660—and who died in 1731, is now principally known for *Robinson Crusoe*, which was first published in 1719, and has since become one of the most widely printed and reprinted and translated novels in the western world.

Defoe, who was born named Foe and added De later to appear patrician, was born of a poor family; his father was a Dissenter from the Church of England and Defoe was unable to attend prestigious schools of the time. Instead he studied to become a Presbyterian minister, yet quickly realized that was a mistake and became a businessman in London, specializing in woolen goods, and wine. He bought a country estate and a ship, although he was often in debt.

He also became a politician; he believed in religious freedom and published pamphlets advocating against James Stuart, the King. He subsequently was sentenced to three days in a pillory based on his political satires; after being released was sentenced to Newgate Prison. His subsequent release from prison was brokered by Robert Harley, the first Earl of Oxford and his colleagues. Harley also paid for some of Defoe's debts. In exchange, Defoe became an intelligence agent for the Tory party.

He was probably five when the great Plague of 1665-1666 killed an estimated 100,000 residents of London, nearly a full quarter of the population within 18 months.

Most historians credit him with publication of 318 to 545 titles to his name: satirical poems; political tracts, nonfiction and novels. He was also said to have used as many as 198 pen-names, a common practice at the time. But in 1692, he was arrested for debts of 700 British pounds, and may have faced total debts of 1,700 pounds. He was forced to declare bankruptcy.

Chronology of his most important works include:

- *The Storm*, 1704, nonfiction. This has been called the first work of modern journalism; a detailed account of a week-long storm that hit London November 24, 1703 and reached its apex the night of November 26-27. Within a week Defoe placed advertisements in newspapers asking readers to submit personal accounts; about 60 were selected by Defoe for the book. This was an innovative method for its time before journalism that relied on first-person accounts.

- *The Adventures of Robinson Crusoe*, 1719, novel. Published when Defoe was in his late 50s, this is the adventures of Crusoe who was shipwrecked on a desert island for 28 years. This is widely assumed to be based on the life of Alexander Selkirk, who spent four years stranded in the Juan Fernandez Islands, a sparsely populated archipelago in the south Pacific owned by Chile. The island Selkirk lived on, "Mas a Tierra," ("Closer to Land") was renamed Robinson Crusoe Island in 1966, yet the book doesn't completely match Selkirk's adventures; the rest, we assume, perhaps Defoe's imagination.

Robinson Crusoe lives on today: the 2000 film *Cast Away* starring Tom Hanks, is clearly a homage to *Crusoe* and *The Man Who Changed His Skin*, (Thomas Fensch, 2011) has a major twentieth century real-life reference to Crusoe.

Howard Griffin, a native of Fort Worth, enlisted in the Army Air Force during World War Two, when the Air Force was not yet a separate division of the U.S. armed forces. Griffin was sent to an island somewhere north of Guadalcanal, acting as a spotter for incoming Japanese aircraft.

He discovered the sole tribe on the island had no written language; Griffin was able to translate it phonetically for the first time.

Griffin was unable to navigate jungle trails without a five-year-old native boy to guide him. He later wrote a novel about his island experiences, *Nuni* (which he translated as *World*), but ruefully admitted that the native islanders were more humane than those he pictured in his novel.

Nuni (and Griffin's experiences) are Robinson Crusoe come to life, Fensch writes.

Griffin lost his eyesight during World War Two because of a bomb blast by Japanese bombers; he was blind for 10 years (and expected to be blind for the rest of his life). Suddenly, after ten years, and for no reason anyone has ever satisfactorily explained, *his eyesight came back.*

He then decided he needed to know what it was like to be black. He dyed his skin black and traveled throughout the south during the pre-Civil Rights years. His subsequent book, ***Black Like Me***, first published in 1961, became an instant American classic and has remained so to this day.

- *Moll Flanders,* 1722, novel, perhaps the second most recognized book in the Defoe canon. Published in the same year as *A Journal of the Plague Year, Moll Flanders* is a picaresque novel. (It was published anonymously; Defoe was confirmed as the author by bookseller Francis Noble in 1770, after Defoe's death in 1731.)

  The definition of picaresque is: "episodic fiction dealing with the adventures of a rough and dishonest but appealing hero," and the book is exactly that.

  It was based—in part—on the life of one Moll King, a prisoner Defoe met in Newgate prison. The full title is:

> The fortunes and misfortunes
> of the famous Moll Flanders,
> Who was born in Newgate,
> and during a life of continu'd Variety for Threescore years,
> Besides her Childhood,
> was Twelve Year a Whore, five times a Wife
> (whereof once to her own Brother)
> Twelve Years a Thief,
> Eight Year a Transported Felon in Virginia,
> At last grew Rich, liv'd Honest,
> and died a Penitent,
> Written from her own memorandums.

The plot? Her mother is a convict in Newgate Prison and gets a reprieve by "pleading her belly," a reference to the postponement of executing pregnant women. The mother is eventually transported to the colonial United States and Moll Flanders (not her birth name, she admits), is raised until three by a kindly foster mother. She then gets attached to a household as a servant; the household has two sons and she is loved by both. The older convinced her to "act like they are married" in bed. Unwilling to marry him, she agrees to marry the younger brother. After five years she is widowed. Leaving her children in the hands of in-laws, she passes herself off as a widow with a fortune, to attract a man who would marry her and provide her with security.

The first time she does this, her "gentleman-tradesman" spendthrift husband (Defoe himself?) goes bankrupt and flees to Europe, leaving her to try and forget him. (They had one child "which was buried.") She then meets a man who takes her to the Virginia Colony and introduces her to his Mother. After three children with that man (one child dies) she learned that her mother-in-law is actually her own biological mother (what are the odds of that?), and her husband is actually her half-brother. She dissolves her marriage, but continues to live with her half-brother for three years, then returns to England. She then meets a man in Bath and eventually becomes a "kept woman," in London. They have three children together; only one lives.

And on and on and on. She becomes a thief, and with her femininity and beauty and artfulness—and her innate hardheartedness—she is successful (for a time), but is eventually caught. She avoids prison by being transported (again) to the colonies, where she learns her Mother has bequeathed her a plantation. Eventually she (and another husband) return to England and, at 69, she has some sort of spiritual epiphany and repents for the "wicked lives we have lived."

The sum total of her "wicked life":

Official marriages:	5
Unofficial marriages:	1
Acted as mistress:	1
Husband deaths:	3

Child deaths:	5
Children accounted for:	7
Children unaccounted for:	2
Children she sees again:	1

Throughout the novel three topics emerge: incest; wickedness—her habitual sins of marriage, affairs and thievery and, finally, redemption—and her own sense of women as agents of their own wealth.

If Defoe's stilted prose of nearly 300 years ago is revised, is this appropriate today? Surely so.

It has been:

- Made into a 1965 film, titled "The Amorous Adventures of Moll Flanders," starring Kim Novak and Angela Landsbury;
- A two-part 1975 BBC adaptation starring Julia Foster as Moll Flanders;
- A 1996 American adaptation "Moll Flanders," starring Robin Wright Penn. Only the name and the main character appear extant; most of the other elements of the original novel are missing;
- A second British adaptation on ITV, in 1996, "The Fortunes and Misfortunes of Moll Flanders," with Alex Kinston as Moll Flanders and Daniel Craig, he of James Bond fame, as Jemmy.

... and there have been other treatments, including radio and stage plays.

There are still a wide variety of editions of *Moll Flanders* available and an equally wide variety of criticism and critical studies of interest.

The Bubonic Plague

Bubonic Plague is one of three types of plague, caused by the plague bacterium *Yersinia pestis*. One to seven days after exposure to the bacteria, flu-like symptoms develop; fever, vomiting, headaches, as well as swollen and painful lymph nodes where the bacteria entered the skin.

Bubonic Plague is transmitted to humans by infected fleas from small animals, usually rats, as a vehicle to jump from animals to humans, an idea not known during the London Plague years of 1665-1666. Vaccines are now available for plague victims including streptomycin, gentamicin and doxycycline, clearly not available during those years. With suitable treatment death rates are now 1-15 percent of those infected.

During the Great Plague, without suitable medical treatment, death rates were 30 to 90 percent, with death usually within 10 days.

The word *Bubonic* comes from a Greek word meaning *groin*.

The flea is a parasite on house and field rats and seeks new prey (as do other plagues, to be cited later), when the host rodent dies. Rats were a major factor of Bubonic Plague because of their close association with humans.

Symptoms include:

- Chills;
- High fever (102.2 F);
- Muscle cramps;
- Seizures;
- Smooth painful lymph node swelling, called a *Bubo*, commonly found the groin, but may occur elsewhere, in the armpits or neck or near where the flea bite occurred;
- Gangrene.

Other symptoms may include:

- Heavy breathing;
- Continuing vomiting (seen in other plagues, also cited later);
- Aching limbs;
- Extreme pain caused by decomposing skin while the victim is still alive;
- Spleen inflation;
- *Lenticulae* (black dots scattered throughout the body);
- Delirium;
- Organ failure;
- Coma …

… and death.

Victims of Bubonic Plague should be given medical treatment immediately; antibiotics within 24 hours. Other treatments now include oxygen, intravenous fluids and respiratory support. People who had contact with Bubonic Plague victims should also receive antibiotics, including streptomycin.

The Bubonic Plague of London, 1665-1666

A bright comet was seen the sky in late 1664, and people in London were fearful—wondering what evil event it predicted.

> London at that time was a city of about 448 acres surrounded by a city wall that had originally been built to keep out raiding bands, and, in the south by the River Thames. There were gates in the wall at Ludgate, Newgate, Aldersgate, Crippengate, Moorgate, Bishopgate and Aldgate, and the Thames was crossable at London Bridge. In the poorer parts of the city, filled with overcrowded tenements and garrets, hygiene was

impossible to maintain. There was no sanitation, and open drains flowed along the center of winding streets. The cobbles were slippery with animal droppings, rubbish and the slops thrown out of the houses; they were muddy and buzzing with flies in the summer and awash with sewage in winter. The City Corporation employed "rakers" to remove the worst of the filth, and it was transported to mounds outside the walls, where it accumulated and continued to decompose. The stench was overwhelming, and people walked around with handkerchiefs or nosegays pressed against their nostrils.

Outside the city, shanty towns with wooden shacks and no sanitation had sprung up, providing homes for the craftsmen and tradespeople who had flocked to the already overcrowded city. The government had tried to limit the development of these "suburbs," but failed.

Over a quarter of a million people lived in them.

And when some well-to-do residents of London fled the city, immigrants took over their homes and crowded into them, turning them into tenements; these homes became rat-infested slums.

And then the 100,000 people died—almost one quarter of all London's population—all in 18 months.

DEFOE'S *A JOURNAL OF THE PLAGUE YEAR*

It was against this background of filth, disease, illnesses—fear of the unknown—and death—that Defoe wrote *A Journal of the Plague Year*, attempting *verisimilitude*, i.e., *the art of—or recreating—reality*. Readers unfamiliar with Defoe might rightly assume he witnessed the plague year as an adult, little realizing he was five years old at the time and had to recreate the entire narrative.

The book was published under the initials H.F., his uncle Henry Poe, who *had* kept journals. Defoe apparently used those journals as a variation of the nonfiction techniques he used in *The Storm*.

The original 1722 edition of *A Journal of the Plague Year* appears in the following pages.

What do we see in Defoe's narrative?

(Some critics have called it historical fiction, others have described it as historical nonfiction, but it does present a comprehensive picture of those times).

- How he "observes" that those who could, fled the city leaving only, he assumes, government workers behind:

  … the number of people there were extremely lessened by so great a multitude having been gone into the country; and even all this month July they continue to flee, though not in such magnitudes as formerly. In August, indeed, they fled in such a manner that I began to think there would be really none but magistrates and servants left in the city.

- He paints the Plague with a remarkable image:

  Death was before their eyes and everybody began to think of their graves, not of mirth and diversion.

- Some London residents gave up religion and pursued other solutions:

  …so they were as mad upon their running after quacks and mountebanks, and every practicing old woman, for medicines and remedies; storing themselves with such multitudes of pills, potions and preservatives, as they were called, that they not only spent their money but even poisoned themselves beforehand for fear of the poison of the infection; and prepared their bodies for the plague, instead of preserving them against it.

Quacks and mountebanks …
… In the third week of April, 2020, then President Donald Trump suggested, in a televised appearance, that bleach might be the solution to the Covid-19 Plague: "I see the disinfectant that knocks it out in a minute, one minute, and is there a way we could do something like that by injection inside, or almost a cleaning ?"

In an article in *The Washington Post*, April 24, 2020, with the headline "Trump claims controversial comment about injecting disinfectants was 'sarcastic.'" His question "immediately spurred doctors, lawmakers and the makers of Lysol to respond with incredulity and warnings against injecting or otherwise ingesting disinfectants, which are highly toxic."

He had to "walk back," as they say now, the idea, claiming he was being sarcastic. "I was asking a question sarcastically to reporters like you to see what would happen." *What would happen* was: hospitals reported people came in for treatment who had injected or swallowed bleach as their President suggested. Replays of the videotape showed not a whit of sarcasm in his voice or manner during the original segment. Dr. Deborah Brix, one of Trump's senior medical advisors, was said to be mortified at his original bleach statement.

Did he ever really apologize for misleading the American public about bleach? It is said that Donald John Trump has never apologized for anything—ever.

And ... a few weeks earlier, he also suggested the use of the anti-malarial drug hydroxychloroquine as a Covid-19 remedy. There was—at best—scant evidence that it was a treatment solution to Covid-19.

A headline in *The New York Times* April 5, 2020 stated:

> Ignoring Expert Opinion, Trump Again
> Promotes Use of Hydroxychloroquine

In that article he said he was speaking "on gut instinct," but added, "what do I know? I'm not a doctor."

Indeed.

But some skeptics—or cynics—on the internet observed that he encouraged the use of hydroxychloroquine so often he must have had stock in the company that made it.

Defoe also writes about the efforts of the Lord Mayor of London to help those who needed help:

> I mean the diseased poor and in particular (he) ordered the College of Physicians to pubhsh directions for cheap remedies for the poor. In all circumstance of the distemper. This, indeed, was one of the most charitable and judicious things that could have been done at that time, for this drove the people from haunting the doors of every dispenser of pills. And from taking down blindly and without consideration poison for physic and death instead of life.

Defoe imagines how the Plague might be seen under a microscope:

> ... of strange, monstrous, and frightful shapes, such as dragons, snakes, serpents and devils, horrible to behold. But this I

very much question the truth of, and we had no microscopes at that time, as I remember, to make the experiment with.

Strange, monstrous and frightful shapes …

… jump forward to October 13, 1976 to the first photograph even taken of a virus particle with a "shepherd's crook": horrifically deadly Ebola, the photograph on page 81 of Richard Preston's *The Hot Zone*.

Defoe paints several remarkable word portraits in his narrative:

At length the man of the house came to the door; he had on his breeches or drawers, and a yellow flannel waistcoat, no stockings, a pair of slipped-shoes, a white cap on his head, and, as the young man said, "death in his face."

… and …

But it was also true that all the people who thus left the land and live-in board the ships were not entirely safe from the infection, for many died and some were thrown overboard into the river, some in coffins and some, as I heard, without coffins, whose bodies were seen sometimes to drive up and down with the tide in the river.

In his original narrative *drive* might well have been *drift*; there have been television clips during the Covid-19 Plague of corpses floating in rivers in India.

And, in another memorable phrase, Defoe writes, that at one point, there were seven or eight thousand deaths per week:

... they died by heaps and were buried by heaps.

- Throughout the narrative Defoe refers to obvious lesions of Bubonic Plague on the human body—for the lack of a better word during his time—as *tokens*.

In the internet article, "What Can Daniel Defoe's 'Plague Year' Teach Us About Coronavirus," Eliott Grover writes:

> The most aggressive measure taken by the city of London in 1665 was forcing all infected individuals to be locked in their homes with their families, even if their family members weren't sick. This "had very great inconveniences in it, and some that were very tragical," Defoe acknowledges, "but was authorized by a law, it had the public good in view as the end chiefly aimed at, and all the private injuries that were done by putting it in execution must be put to account of the public benefit."
>
> Many people died because they were confined with sick relatives but many more were saved by keeping potentially infected people off the streets.

Daniel Defoe died April 23, i731, probably from a stroke, and while quite possibly hiding from a bill collector.

The first, 1722, edition of Daniel Defoe's *A Journal of the Plague Year* contains the following sub-title, in a variety of type fonts (type styles), large and smaller, popular at that time:

BEING
Observations or Memorials,
Of the Most Remarkable
OCCURRENCES,
As well
PUBLICK as PRIVATE,
Which happened in
LONDON
During the last
GREAT VISITATION
In 1665.

Written by a CITIZEN who continued all the while in London. Never made publick before.

Is *A Journal of the Plague Year* historical fiction or historical non-fiction, or is it nascent journalism, as was Defoe's *The Storm*, published 1704?

It may well be a combination of all three; but essentially *Plague Year* is the longest and most detailed record we have of the 1665 plague that decimated London. Although there are wide variety of editions of *Plague Year* available, we are including it here in its entirety.

A Journal of the Plague Year
by Daniel Defoe

It was about the beginning of September, 1664, that I, among the rest of my neighbours, heard in ordinary discourse that the plague was returned again in Holland; for it had been very violent there, and particularly at Amsterdam and Rotterdam, in the year 1663, whither, they say, it was brought, some said from Italy, others from the Levant, among some goods which were brought home by their Turkey fleet; others said it was brought from Candia; others from Cyprus. It mattered not from whence it came; but all agreed it was come into Holland again.

We had no such thing as printed newspapers in those days to spread rumours and reports of things, and to improve them by the invention of men, as I have lived to see practised since. But such things as these were gathered from the letters of merchants and others who corresponded abroad, and from them was handed about by word of mouth only; so that things did not spread instantly over the whole nation, as they do now. But it seems that the Government had a true account of it, and several councils were held about ways to prevent its coming over; but all was kept very private. Hence it was that this rumour died off again, and people began to forget it as a thing we were very little concerned in, and that we hoped was not true; till the latter end of November or the beginning of December 1664 when two men, said to be Frenchmen, died of the plague in Long Acre, or rather at the upper end of Drury Lane. The family they were in endeavoured to conceal it as much as possible, but as it had gotten some vent in the discourse of the neighbourhood, the Secretaries of State got knowledge of it; and concerning themselves to inquire about it, in order to be certain of the truth, two physicians and a surgeon were ordered to go to the house and make inspection. This they did; and finding evident tokens of the sickness upon both the bodies that were dead, they gave their opinions publicly that they died of the plague. Whereupon it was given in to the parish clerk, and he also returned them to the Hall; and it was printed in the weekly bill of mortality in the usual manner, thus:—

Plague, 2. Parishes infected, 1.

The people showed a great concern at this, and began to be alarmed all over the town, and the more, because in the last week in December 1664 another man died in the same house, and of the same distemper. And then we were easy again for about *six* weeks, when none having died with any marks of infection, it was said the distemper was gone; but after that, I think it was about the 12th of February, another died in another house, but in the same parish and in the same manner.

This turned the people's eyes pretty much towards that end of the town, and the weekly bills showing an increase of burials in St Giles's parish more than usual, it began to be suspected that the plague was among the people at that end of the town, and that many had died of it, though they had taken care to keep it as much from the knowledge of the public as possible. This possessed the heads of the people very much, and few cared to go through Drury Lane, or the other streets suspected, unless they had extraordinary business that obliged them to it

This increase of the bills stood thus: the usual number of burials in a week, in the parishes of St Giles-in-the-Fields and St Andrew's, Holbom, were from twelve to seventeen or nineteen each, few more or less; but from the time that the plague first began in St Giles's parish, it was observed that the ordinary burials increased in number considerably. For example:—

From December 27 to January 3 { St Giles's 16
{ St Andrew's 17
" January 3 " " 10 { St Giles's 12
{ St Andrew's 25 " January 10 " " 17 { St Giles's 18
{ St Andrew's 28 " January 17 " " 24 { St Giles's 23
{ St Andrew's 16
" January 24 " " 31 { St Giles's 24
{ St Andrew's 15 " January 30 " February 7 { St Giles's 21
{ St Andrew's 23
" February 7 " " 14 { St Giles's 24

Whereof one of the plague.

The like increase of the bills was observed in the parishes of St Bride's, adjoining on one side of Holbom parish, and in the parish of St James, Clerkenwell, adjoining on the other side of Holbom; in both which parishes the usual numbers that died weekly were from four to six or eight, whereas at that time they were increased as follows:—

From December 20 to December 27 { St Bride's 0
{ St James's 8
December 27 to January 3 { St Bride's 6
{ St James's 9 " January 3 " " 10 { St Bride's 11
{ St James's 7 " January 10 " " 17 { St Bride's 12
{ St James's 9 " January 17 " " 24 { St Bride's 9
{ St James's 15
" January 24 " " 31 { St Bride's 8
{ St James's 12 "January 31 " February 7 { St Bride's 13
{ St James's 5
" February 7 " " 14 { St Bride's 12
{ St James's 6

Besides this, it was observed with great uneasiness by the people that the weekly bills in general increased very much during these weeks, although it was at a time of the year when usually the bills are very moderate.

The usual number of burials within the bills of mortality for a week was from about 240 or thereabouts to 300. The last was esteemed a pretty high bill; but after this we found the bills successively increasing as follows;—

Buried. Increased.
    December the 20th to the 27th 291 ...
    " 27th " 3rd January 349 58
    January the 3rd " 10th " 394 45
    " 10th " 17th " 415 21
    " 17th " 24th " 474 59

This last bill was really frightful, being a higher number than had been known to have been buried in one week since the preceding visitation of 1656.

However, all this went off again, and the weather proving cold, and the frost, which began in December, still continuing very severe even till near the end of February, attended with sharp though moderate winds, the bills decreased again, and the city grew healthy, and everybody began to look upon the danger as good as over; only that still the burials in St Giles's continued high. From the beginning of April especially they stood at twenty-five each week, till the week from the 18th to the 25th, when there was buried in St Giles's parish thirty, whereof two of the plague and eight of the spotted-fever, which was looked upon as the same thing; likewise the number that died of the spotted-fever in the whole increased, being eight the week before, and twelve the week above-named.

This alarmed us all again, and terrible apprehensions were among the people, especially the weather being now changed and growing warm, and the summer being at hand. However, the next week there seemed to be some hopes again; the bills were low, the number of the dead in all was but 388, there was none of the plague, and but four of the spotted-fever.

But the following week it returned again, and the distemper was spread into two or three other parishes, viz., St Andrew's, Holbom; St Clement Danes; and, to the great affliction of the city, one died within the walls, in the parish of St Mary Woolchurch, that is to say, in Bearbinder Lane, near Stocks Market; in all there were nine of the plague and six. of the spotted-fever. It was, however, upon inquiry found that this Frenchman who died in Bearbinder Lane was one who, having lived in Long Acre, near the infected houses, had removed for fear of the distemper, not knowing that he was already infected.

This was the beginning of May, yet the weather was temperate, variable, and cool enough, and people had still some hopes. That which encouraged them was that the city was healthy: the whole ninety-seven parishes buried but fifty-four, and we began to hope that, as it was chiefly among the people at that end of the town, it might go no farther; and the rather, because the next week, which was from the 9th of May to the 16th, there died but three, of which not one within the whole city or liberties; and St Andrew's buried but fifteen, which was very low. 'Tis true St Giles's buried two-and-thirty, but still, as there was but one of the plague, people began to be easy. The whole bill also was very low, for the week before the bill was but 347, and the week above mentioned but 343. We continued in these hopes for a few days, but it was but for a few, for the people were no more to be deceived thus; they searched the houses and found that the plague was really spread every way, and that many died of it every day. So that now all our extenuations abated, and it was no more to be concealed; nay, it quickly appeared that the infection had spread itself beyond all hopes of abatement, that in the parish of St Giles it was gotten into several streets, and several families lay all sick together; and, accordingly, in the weekly bill for the next week the thing began to show itself. There was indeed but fourteen set down of the plague, but this was all knavery and collusion, for in St Giles's parish they buried forty in all, whereof it was certain most of them died of the plague, though they were set down of other distempers; and though the number of all the burials were not increased above thiily-two, and the whole bill being but 385, yet there was fourteen of the spotted-fever, as well as fourteen of the plague; and we took it for granted upon the whole that there were fifty died that week of the plague.

The next bill was from the 23rd of May to the 30th, when the number of the plague was seventeen. But the burials in St Giles's were fifty-three—a frightful number!—of whom they set down but nine of the plague; but on an examination more strictly by the justices of peace, and at the Lord Mayor's request, it was found there were twenty more who were really dead of the plague in that parish, but had been set down of the spotted-fever or other distempers, besides others concealed.

But those were trifling things to what followed immediately after; for now the weather set in hot, and from the first week in June the infection spread in a dreadful manner, and the bills rose high; the articles of the fever, spotted-fever, and teeth began to swell; for all that could conceal their distempers did it, to prevent their neighbours shunning and refusing to converse with them, and also to prevent authority shutting up their houses; which, though it was not yet practised, yet was threatened, and people were extremely terrified at the thoughts of it.

The second week in June, the parish of St Giles, where still the weight of the infection lay, buried 120, whereof though the bills said but sixty-eight of the plague, everybody said there had been 100 at least, calculating it from the usual number of funerals in that parish, as above.

Till this week the city continued free, there having never any died, except that one Frenchman whom I mentioned before, within the whole ninety-seven parishes. Now there died four within the city, one in Wood Street, one in Fenchurch Street, and two in Crooked Lane. Southwark was entirely free, having not one yet died on that side of the water.

I lived without Aldgate, about midway between Aldgate Church and Whitechappel Bars, on the left hand or north side of the street; and as the distemper had not reached to that side of the city, our neighbourhood continued very easy. But at the other end of the town their consternation was very great: and the richer sort of people, especially the nobility and gentry from the west part of the city, thronged out of town with their families and servants in an unusual manner; and this was more particularly seen in Whitechappel; that is to say, the Broad Street where I lived; indeed, nothing was to be seen but waggons and carts, with goods, women, servants, children, &c.; coaches filled with people of the better sort and horsemen attending them, and all hurrying away; then empty waggons and carts appeared, and spare horses with servants, who, it was apparent, were returning or sent from the countries to fetch more people; besides innumerable numbers of men on horseback, some alone, others with servants, and, generally speaking, all loaded with baggage and fitted out for travelling, as anyone might perceive by their appearance.

This was a very terrible and melancholy thing to see, and as it was a sight which I could not but look on from morning to night (for indeed there was nothing else of moment to be seen), it filled me with very serious thoughts of the misery that was coming upon the city, and the unhappy condition of those that would be left in it.

This hurry of the people was such for some weeks that there was no getting at the Loiri Mayor's door without exceeding difficulty; there were such pressing and crowding there to get passes and certificates of health for such as travelled abroad, for without these there was no being admitted to pass through the towns upon the road, or to lodge in any inn. Now, as there had none died in the city for all this time, my Lord Mayor gave certificates of health without any difficulty to all those who lived in the ninety-seven parishes, and to those within the liberties too for a while.

This hurry, I say, continued some weeks, that is to say, all the month of May and June, and the more because it was rumoured that an order of the Government was to be issued out to place turnpikes and barriers on the road to prevent people travelling, and that the towns on the road would not suffer people from London to pass for fear of bringing the infection along with them, though neither of these rumours had any foundation but in the imagination, especially at-first.

I now began to consider seriously with myself concerning my own case, and how I should dispose of myself; that is to say, whether I should resolve to stay in London or shut up my house and flee, as many of my neighbours did. I have set this particular down so fully, because I know not but it may be of moment to those who come after me, if they come to be brought to the same distress, and to the same manner of making their choice; and therefore I desire this account may pass with them rather for a direction to themselves to act by than a history of my actings, seeing it may not be of one farthing value to them to note what became of me.

I had two important things before me: the one was the carrying on my business and shop, which was considerable, and in which was embarked all my effects in the world; and the other was the preservation of my life in so dismal a calamity as I saw apparently was coming upon the whole city, and which, however great it was, my fears perhaps, as well as other people's, represented to be much greater than it could be.

The first consideration was of great moment to me; my trade was a saddler, and as my dealings were chiefly not by a shop or chance trade, but among the merchants trading to the English colonies in America, so my effects lay very much in the hands of such. I was a single man, 'tis true, but I had a family of servants whom I kept at my business; had a house, shop, and warehouses filled with goods; and, in short, to leave them all as things in such a case must be left (that is to say, without any overseer or person fit to be trusted with them), had been to hazard the loss not only of my trade, but of my goods, and indeed of all I had in the world.

I had an elder brother at the same time in London, and not many years before come over from Portugal: and advising with him, his answer was in three words, the same that was given in another case quite different, viz., 'Master, save thyself.' In a word, he was for my retiring into the country, as he resolved to do himself with his family; telling me what he had, it seems, heard abroad, that the best preparation for the plague was to run away from it. As to my argument of losing my trade, my goods, or debts, he quite confuted me. He told me the same thing which I argued for my staying, viz., that I would trust God with my safety and health, was the strongest repulse to my pretensions of losing my trade and my goods; 'for', says he, 'is it not as reasonable that you should trust God with the chance or risk of losing your trade, as that you should stay in so eminent a point of danger, and trust Him with your life?'

I could not argue that I was in any strait as to a place where to go, having several friends and relations in Northamptonshire, whence our family first came from; and particularly, I had an only sister in Lincolnshire, very willing to receive and entertain me.

My brother, who had already sent his wife and two children into Bedfordshire, and resolved to follow them, pressed my going very earnestly; and I had once resolved to comply with his desires, but at that time could get no horse; for though it is true all the people did not go out of the city of London, yet I may venture to say that in a manner all the horses did; for there was hardly a horse to be bought or hired in the whole city for some weeks. Once I resolved to travel on foot with one servant, and, as many did, lie at no inn, but carry a soldier's tent with us, and so lie in the fields, the weather being very warm, and no danger from taking cold. I say, as many did, because several did so at last, especially those who had been in the armies in the war which had not been many years past; and I must needs say that, speaking of second causes, had most of the people that travelled done so, the plague had not been carried into so many country towns and houses as it was, to the great damage, and indeed to the ruin, of abundance of people.

But then my servant, whom I had intended to take down with me, deceived me; and being frighted at the increase of the distemper, and not knowing when I should go, he took other measures, and left me, so I was put off for that time; and, one way or other, I always found that to appoint to go away was always crossed by some accident or other, so as to disappoint and put it off again; and this brings in a story which otherwise might be thought a needless digression, viz., about these disappointments being from Heaven.

I mention this story also as the best method I can advise any person to take in such a case, especially if he be one that makes conscience of his duty, and would be directed what to do in it, namely, that he should keep his eye upon the particular providences which occur at that time, and look upon them complexly, as they regard one another, and as all together regard the question before him: and then, I think, he may safely take them for intimations from Heaven of what is his unquestioned duty to do in such a case; I mean as to going away from or staying in the place where we dwell, when visited with an infectious distemper.

It came very warmly into my mind one morning, as I was musing on this particular thing, that as nothing attended us without the direction or permission of Divine Power, so these disappointments must have something in them extraordinary; and I ought to consider whether it did not evidently point out, or intimate to me, that it was the will of Heaven I should not go. It immediately followed in my thoughts, that if it really was from God that I should stay. He was able effectually to preserve me in the midst of all the death and danger that would surround me; and that if I attempted to secure myself by fleeing from my habitation, and acted contrary to these intimations, which I believe to be Divine, it was a kind of flying from God, and that He could cause His justice to overtake me when and where He thought fit.

These thoughts quite turned my resolutions again, and when I came to discourse with my brother again I told him that I inclined to stay and take my lot in that station in which God had placed me, and that it seemed to be made more especially my duty, on the account of what I have said.

My brother, though a very religious man himself, laughed at all I had suggested about its being an intimation from Heaven, and told me several stories of such foolhardy people, as he called them, as I was; that I ought indeed to submit to it as a work of Heaven if I had been any way disabled by distempers or diseases, and that then not being able to go, I ought to acquiesce in the direction of Him, who, having been my Maker, had an undisputed right of sovereignty in disposing of me, and that then there had been no difficulty to determine which was the call of His providence and which was not; but that I should take it as an intimation from Heaven that I should not go out of town, only because I could not hire a horse to go, or my fellow was run away that was to attend me, was ridiculous, since at the time I had my health and hmbs, and other servants, and might with ease travel a day or two on foot, and having a good certificate of being in perfect health, might either hire a horse or take post on the road, as I thought fit.

Then he proceeded to tell me of the mischievous consequences which attended the presumption of the Turks and Mahometans in Asia and in other places where he had been (for my brother, being a merchant, was a few years before, as I have already observed, returned from abroad, coming last from Lisbon), and how, presuming upon their professed predestinating notions, and of every man's end being predetermined and unalterably beforehand decreed, they would go unconcerned into infected places and converse with infected persons, by which means they died at the rate of ten or fifteen thousand a week, whereas the Europeans or Christian merchants, who kept themselves retired and reserved, generally escaped the contagion.

Upon these arguments my brother changed my resolutions again, and I began to resolve to go, and accordingly made all things ready; for, in short, the infection increased round me, and the bills were risen to almost seven hundred a week, and my brother told me he would venture to stay no longer. I desired him to let me consider of it but till the next day, and I would resolve: and as I had already prepared everything as well as I could as to MY business, and whom to entrust my affairs with, I had little to do but to resolve.

I went home that evening greatly oppressed in my mind, irresolute, and not knowing what to do. I had set the evening wholly apart to consider seriously about it, and was all alone; for already people had, as it were by a general consent, taken up the custom of not going out of doors after sunset; the reasons I shall have occasion to say more of by-and-by.

In the retirement of this evening I endeavoured to resolve, first, what was my duty to do, and I stated the arguments with which my brother had pressed me to go into the country, and I set, against them the strong impressions which I had on my mind for staying; the visible call I seemed to have from the particular circumstance of my calling, and the care due from me for the preservation of my effects, which were, as I might say, my estate; also the intimations which I thought I had from Heaven, that to me signified a kind of direction to venture; and it occurred to me that if I had what I might call a direction to stay, I ought to suppose it contained a promise of being preserved if I obeyed.

This lay close to me, and my mind seemed more and more encouraged to stay than ever, and supported with a secret satisfaction that I should be kept. Add to this, that, turning over the Bible which lay before me, and while my thoughts were more than ordinarily serious upon the question, I cried out, 'Well, I know not what to do; Lord, direct me T and the like; and at that juncture I happened to stop turning over the book at the gist Psalm, and casting my eye on the second verse, I read on to the seventh verse exclusive, and after that included the tenth, as follows: 'I will say of the Lord, He is my refuge and my fortress: my God, in Him will I trust. Surely He shall deliver thee from the snare of the fowler, and from the noisome pestilence. He shall cover thee with His feathers, and under His wings shalt thou trust: His truth shall be thy shield and buckler. Thou shalt not be afraid for the terror by night; nor for the arrow that flieth by day; nor for the pestilence that walketh in darkness; nor for the destruction that wasteth at noonday. A thousand shall fall at thy side, and ten thousand at thy right hand; but it shall not come nigh thee. Only with thine eyes shaft thou behold and see the reward of the wicked. Because thou hast made the Lord, which is my refuge, even the most High, thy habitation; there shall no evil befall thee, neither shall any plague come nigh thy dwelling,' &C.

I scarce need tell the reader that from that moment I resolved that I would stay in the town, and casting myself entirely upon the goodness and protection of the Almighty, would not seek any other shelter whatever; and that, as my times were in His hands. He was as able to keep me in a time of the infection as in a time of health; and if He did not think fit to deliver me, still I was in His hands, and it was meet He should do with me as should seem good to Him.

With this resolution I went to bed; and I was further confirmed in it the next day by the woman being taken ill with whom I had intended to entrust my house and all my affairs. But I had a further obligation laid on me on the same side, for the next day I found myself very much out of order also, so that if I would have gone away, I could not, and I continued ill three or four days, and this entirely determined my stay; so I took my leave of my brother, who went away to Dorking, in Surrey, and afterwards fetched a round farther into Buckinghamshire or Bedfordshire, to a retreat he had found out there for his family.

It was a very ill time to be sick in, for if any one complained, it was immediately said he had the plague; and though I had indeed no symptom of that distemper, yet being very ill, both in my head and in my stomach, I was not without apprehension that I really was infected; but in about three days I grew better; the third night I rested well, sweated a little, and was much refreshed. The apprehensions of its being the infection went also quite away with my illness, and I went about my business as usual.

These things, however, put off all my thoughts of going into the country; and my brother also being gone, I had no more debate either with him or with myself on that subject.

It was now mid-July, and the plague, which had chiefly raged at the other end of the town, and, as I said before, in the parishes of St Giles, St Andrew's, Holbom, and towards Westminster, began to now come eastward towards the part where I lived. It was to be observed, indeed, that it did not come straight on towards us; for the city, that is to say, within the walls, was indifferently healthy still; nor was it got then very much over the water into Southwark; for though there died that week 1268 of all distempers, whereof it might be supposed above 600 died of the plague, yet there was but twenty-eight in the whole city, within the walls, and but nineteen in Southwark, Lambeth parish included; whereas in the parishes of St Giles and St Martin-in-the-Fields alone there died 421.

But we perceived the infection kept chiefly in the out-parishes, which being very populous, and fuller also of poor, the distemper found more to prey upon than in the city, as I shall observe afterwards. We perceived, I say, the distemper to draw our way, viz., by the parishes of Clarkenwell, Cripplegate, Shoreditch, and Bishopsgate; which last two parishes joining to Aldgate, Whitechappel, and Stepney, the infection came at length to spread its utmost rage and violence in those parts, even when it abated at the western parishes where it began.

It was very strange to observe that in this particular week, from the 4th to the 11th of July, when, as I have observed, there died near 400 of the plague in the two parishes of St Martin and St Giles-in-the-Fields only, there died in the parish of Aldgate but four, in the parish of Whitechappel three, in the parish of Stepney but one.

Likewise in the next week, from the 11th of July to the 18th, when the week's bill was 1761, yet there died no more of the plague, on the whole Southwark side of the water, than sixteen. But this face of things soon changed, and it began to thicken in Cripplegate parish especially, and in Clarkenwell; so that by the second week in August, Cripplegate parish alone buried 886, and Clarkenwell 155. Of the first, 850 might well be reckoned to die of the plague; and of the last, the bill itself said 145 were of the plague.

During the month of July, and while, as I have observed, our part of the town seemed to be spared in comparison of the west part, I went ordinarily about the streets, as my business required, and particularly went generally once in a day, or in two days, into the city, to my brother's house, which he had given me charge of, and to see if it was safe; and having the key in my pocket, I used to go into the house, and over most of the rooms, to see that all was well; for though it be something wonderful to tell, that any should have hearts so hardened in the midst of such a calamity as to rob and steal, yet certain it is that all sorts of villainies, and even levities and debaucheries, were then practised in the town as openly as ever—I will not say quite as frequently, because the numbers of people were many ways lessened.

But the city itself began now to be visited too, I mean within the walls; but the number of people there were indeed extremely lessened by so great a multitude having been gone into the country; and even all this month of July they continued to flee, though not in such multitudes as formerly. In August, indeed, they fled in such a manner that I began to think there would be really none but magistrates and servants left in the city.

As they fled now out of the city, so I should observe that the Court removed early, viz., in the month of June, and went to Oxford, where it pleased God to preserve them; and the distemper did not, as I heard of, so much as touch them, for which I cannot say that I ever saw they showed any great token of thankfulness, and hardly anything of reformation, though they did not want being told that their crying vices might without breach of charity be said to have gone far in bringing that terrible judgement upon the whole nation.

The face of London was now indeed strangely altered: I mean the whole mass of buildings, city, liberties, suburbs, Westminster, Southwark, and altogether; for as to the particular part called the city, or within the walls, that was not yet much infected. But in the whole the face of things, I say, was much altered; sorrow and sadness sat upon every face; and though some parts were not yet overwhelmed, yet all looked deeply concerned; and, as we saw it apparently coming on, so every one looked on himself and his family as in the utmost danger. Were it possible to represent those times exactly to those that did not see them, and give the reader due ideas of the horror 'that everywhere presented itself, it must make Just impressions upon their minds and fill them with surprise. London might well be said to be all in tears; the mourners did not go about the streets indeed, for nobody put on black or made a formal dress of mourning for their nearest friends; but the voice of mourners was truly heard in the streets. The shrieks of women and children at the windows and doors of their houses, where their dearest relations were perhaps dying, or just dead, were so frequent to be heard as we passed the streets, that it was enough to pierce the stoutest heart in the world to hear them. Tears and lamentations were seen almost in every house, especially in the first part of the visitation; for towards the latter end men's hearts were hardened, and death was so always before their eyes, that they did not so much concern themselves for the loss of their friends, expecting that themselves should be summoned the next hour.

Business led me out sometimes to the other end of the town, even when the sickness was chiefly there; and as the thing was new to me, as well as to everybody else, it was a most surprising thing to see those streets which were usually so thronged now grown desolate, and so few people to be seen in them, that if I had been a stranger and at a loss for my way, I might sometimes have gone the length of a whole street (I mean of the by-streets), and seen nobody to direct me except watchmen set at the doors of such houses as were shut up, of which I shall speak presently.

One day, being at that part of the town on some special business, curiosity led me to observe things more than usually, and indeed I walked a great way where I had no business. I went up Holbom, and there the street was full of people, but they walked in the middle of the great street, neither on one side or other, because, as I suppose, they would not mingle with anybody that came out of houses, or meet with smells and scent from houses that might be infected.

The Inns of Court were all shut up; nor were very many of the lawyers in the Temple, or Lincoln's Inn, or Gray's Inn, to be seen there. Everybody was at peace; there was no occasion for lawyers; besides, it being in the time of the vacation too, they were generally gone into the country. Whole rows of houses in some places were shut close up, the inhabitants all fled, and only a watchman or two left.

When I speak of rows of houses being shut up, I do not mean shut up by the magistrates, but that great numbers of persons followed the Court, by the necessity of their employments and other dependences; and as others retired, really frighted with the distemper, it was a mere desolating of some of the streets. But the fright was not yet near so great in the city, abstractly so called, and particularly because, though they were at first in a most inexpressible consternation, yet as I have observed that the distemper intermitted often at first, so they were, as it were, alarmed and unalarmed again, and this several times, till it began to be familiar to them; and that even when it appeared Adolent, yet seeing it did not presently spread into the city, or the east and south parts, the people began to take courage, and to be, as I may say, a little hardened. It is true a vast many people fled, as I have observed, yet they were chiefly from the west end of the town, and from that we call the heart of the city: that is to say, among the wealthiest of the people, and such people as were unencumbered with trades and business. But of the rest, the generality stayed, and seemed to abide the worst; so that in the place we calf the Liberties, and in the suburbs, in Southwark, and in the east part, such as Wapping, Ratcliff, Stepney, Rotherhithe, and the like, the people generally stayed, except here and there a few wealthy families, who, as above, did not depend upon their business.

It must not be forgot here that the city and suburbs were prodigiously full of people at the time of this visitation, I mean at the time that it began; for though I have lived to see a further increase, and mighty throngs of people settling in London more than ever, yet we had always a notion that the numbers of people which, the wars being over, the armies disbanded, and the royal family and the monarchy being restored, had flocked to London to settle in business, or to depend upon and attend the Court for rewards of services, preferments, and the like, was such that the town was computed to have in it above a hundred thousand people more than ever it held before; nay, some took upon them to say it had twice as many, because all the ruined families of the royal party flocked hither. All the old soldiers set up trades here, and abundance of families settled here. Again, the Court brought with them a great flux of pride, and new fashions. All people were grown gay and luxurious, and the joy of the Restoration had brought a vast many families to London.

I often thought that as Jerusalem was besieged by the Romans when the Jews were assembled together to celebrate the Passover—by which means an incredible number of people were surprised there who would otherwise have been in other countries—so the plague entered London when an incredible increase of people had happened occasionally, by the particular circumstances above-named. As this conflux of the people to a youthful and gay Court made a great trade in the city, especially in everything that belonged to fashion and finery, so it drew by consequence a great number of workmen, manufacturers, and the like, being mostly poor people who depended upon their labour. And I remember in particular that in a representation to my Lord Mayor of the condition of the poor, it was estimated that there were no less than an hundred thousand riband-weavers in and about the city, the chiefest number of whom lived then in the parishes of Shoreditch, Stepney, Whitechappel, and Bishopsgate, that, namely, about Spitalfields; that is to say, as Spitalfields was then, for it was not so large as now by one fifth part.

By this, however, the number of people in the whole may be judged of; and, indeed, I often wondered that, after the prodigious numbers of people that went away at first, there was yet so great a multitude left as it appeared there was.

But I must go back again to the beginning of this surprising time. While the fears of the people were young, they were increased strangely by several odd accidents which, put altogether, it was really a wonder the whole body of the people did not rise as one man and abandon their dwellings, leaving the place as a space of ground designed by Heaven for an Akeldama, doomed to be destroyed from the face of the earth, and that all that would be found in it would perish with it. I shall name but a few of these things; but sure they were so many, and so many wizards and cunning people propagating them, that I have often wondered there was any (women especially) left behind.

In the first place, a blazing star or comet appeared for several months before the plague, as there did the year after another, a little before the fire. The old women and the phlegmatic hypochondriac part of the other sex, whom I could almost call old women too, remarked (especially afterward, though not till both those judgements were over) that those two comets passed directly over the city, and that so very near the houses that it was plain they imported something peculiar to the city alone; that the comet before the pestilence was of a faint, dull, languid colour, and its motion very heavy. Solemn, and slow; but that the comet before the fire was bright and spariding, or, as others said, flaming, and its motion swift and furious; and that, accordingly, one foretold a heavy judgement, slow but severe, terrible and frightful, as was the plague; but the other foretold a stroke, sudden, swift, and fiery as the conflagration. Nay, so particular some people were, that as they looked upon that comet preceding the fire, they fancied that they not only saw it pass swiftly and fiercely, and could perceive the motion with their eye, but even they heard it; that it made a rushing, mighty noise, fierce and terrible, though at a distance, and but just perceivable.

I saw both these stars, and, I must confess, had so much of the common notion of such things in my head, that I was apt to look upon them as the forerunners and warnings of God's judgements; and especially when, after the plague had followed the first, I yet saw another of the like kind, I could not but say God had not yet sufficiently scourged the city.

But I could not at the same time carry these things to the height that others did, knowing, too, that natural causes are assigned by the astronomers for such things, and that their motions and even their revolutions are calculated, or pretended to be calculated, so that they cannot be so perfectly called the forerunners or foretellers, much less the procurers, of such events as pestilence, war, fire, and the like.

But let my thoughts and the thoughts of the philosophers be, or have been, what they will, these things had a more than ordinary influence upon the minds of the common people, and they had almost universal melancholy apprehensions of some dreadful calamity and judgement coming upon the city; and this principally from the sight of this comet, and the little alarm that was given in December by two people dying at St Giles's, as above.

The apprehensions of the people were likewise strangely increased by the error of the times; in which, I think, the people, from what principle I cannot imagine, were more addicted to prophecies and astrological conjurations, dreams, and old wives' tales than ever they were before or since. Whether this unhappy temper was originally raised by the follies of some people who got money by it—that is to say, by printing predictions and prognostications—I know not; but certain it is, books frighted them terribly, such as Lilly's Almanack, Cadbury's Astrological Predictions, Poor Robin's Almanack, and the like; also several pretended religious books, one entitled. Come out of her, my People, lest you be Partaker of her Plagues; another called. Fair Warning; another, Britain's Remembrancer; and many such, all, or most part of which, foretold, directly or covertly, the ruin of the city. Nay, some were so enthusiastically bold as to run about the streets with their oral predictions, pretending they were sent to preach to the city; and one in particular, who, like Jonah to Nineveh, cried in the streets, 'Yet forty days, and London shall be destroyed.' I will not be positive whether he said yet forty days or yet a few days. Another ran about naked, except a pair of drawers about his waist, crying day and night, like a man that Josephus mentions, who cried, 'Woe to Jerusalem!' a little before the destruction of that city. So this poor naked creature cried, 'Oh, the great and the dreadful God!' and said no more, but repeated those words continually, with a voice and countenance full of horror, a swift pace; and nobody could ever find him to stop or rest, or take any sustenance, at least that ever I could hear of. I met this poor creature several times in the streets, and would have spoken to him, but he would not enter into speech with me or any one else, but held on his dismal cries continually.

These things terrified the people to the last degree, and especially when two or three times, as I have mentioned already, they found one or two in the bills dead of the plague at St Giles's .

Next to these public things were the dreams of old women, or, I should say, the interpretation of old women upon other people's dreams; and these put abundance of people even out of their wits. Some heard voices warning them to be gone, for that there would be such a plague in London, so that the living would not be able to bury the dead. Others saw apparitions in the air; and I must be allowed to say of both, I hope without breach of charity, that they heard voices that never spake, and saw sights that never appeared; but the imagination of the people was really turned wayward and possessed. And no wonder, if they who were poring continually at the clouds saw shapes and figures, representations and appearances, which had nothing in them but air, and vapour. Here they told us they saw a flaming sword held in a hand coming out of a cloud, with a point hanging directly over the city; there they saw hearses and coffins in the air carrying to be buried; and there again, heaps of dead bodies lying unburied, and the like, just as the imagination of the poor terrified people furnished them with matter to work upon.

> So hypochondriac fancies represent
> Ships, armies, battles in the firmament;
> Till steady eyes the exhalations solve.
> And all to its first matter, cloud, resolve.

I could fill this account with the strange relations such people gave every day of what they had seen; and every one was so positive of their having seen what they pretended to see, that there was no contradicting them without breach of friendship, or being accounted rude and unmannerly on the one hand, and profane and impenetrable on the other. One time before the plague was begun (otherwise than as I have said in St Giles's), I think it was in March, seeing a crowd of people in the street, I joined with them to satisfy my curiosity, and found them all staring up into the air to see what a woman told them appeared plain to her, which was an angel clothed in white, with a fiery sword in his hand, waving it or brandishing it over his head. She described every part of the figure to the life, showed them the motion and the form, and the poor people came into it so eagerly, and with so much readiness; 'Yes, I see it all plainly,' says one; 'there's the sword as plain as can be.' Another saw the angel. One saw his very face, and cried out what a glorious creature he was! One saw one thing, and one another. I looked as earnestly as the rest, but perhaps not with so much willingness to be imposed upon; and I said, indeed, that I could see nothing but a white cloud, bright on one side by the shining of the sun upon the other part. The woman endeavoured to show it me, but could not make me confess that I saw it, which, indeed, if I had I must have lied. But the woman, turning upon me, looked in my face, and fancied I laughed, in which her imagination deceived her too, for I really did not laugh, but was very seriously reflecting how the poor people were terrified by the force of their own imagination. However, she turned from me, called me profane fellow, and a scoffer; told me that it was a time of God's anger, and dreadful judgements were approaching, and that despisers such as I should wander and perish.

The people about her seemed disgusted as well as she; and I found there was no persuading them that I did not laugh at them, and that I should be rather mobbed by them than be able to undeceive them. So I left them; and this appearance passed for as real as the blazing star itself.

Another encounter I had in the open day also; and this was in going through a narrow passage from Petty France into Bishopsgate Churchyard, by a row of alms-houses. There are two churchyards to Bishopsgate church or parish; one we go over to pass from the place called Petty France into Bishopsgate Street, coming out just by the church door; the other is on the side of the narrow passage where the alms-houses are on the left; and a dwarf-wall with a palisado on it on the right hand, and the city wall on the other side more to the right.

In this narrow passage stands a man looking through between the palisadoes into the burying-place, and as many people as the narrowness of the passage would admit to stop, without hindering the passage of others, and he was talking mightily eagerly to them, and pointing now to one place, then to another, and affirming that he saw a ghost walking upon such a gravestone there. He described the shape, the posture, and the movement of it so exactly that it was the greatest matter of amazement to him in the world that everybody did not see it as well as he. On a sudden he would cry, 'There it is; now it comes this way.' Then, 'Tis turned back'; till at length he persuaded the people into so firm a belief of it, that one fancied he saw it, and another fancied he saw it; and thus he came every day making a strange hubbub, considering it was in so narrow a passage, till Bishopsgate clock struck eleven, and then the ghost would seem to start, and, as if he were called away, disappeared on a sudden.

I looked earnestly every way, and at the very moment that this man directed, but could not see the least appearance of anything; but so positive was this poor man, that he gave the people the vapours in abundance, and sent them away trembling and frighted, till at length few people that knew of it cared to go through that passage, and hardly anybody by night on any account whatever.

This ghost, as the poor man affirmed, made signs to the houses, and to the ground, and to the people, plainly intimating, or else they so understanding it, that abundance of the people should come to be buried in that churchyard, as indeed happened; but that he saw such aspects I must acknowledge I never believed, nor could I see anything of it myself, though I looked most earnestly to see it, if possible.

These things serve to show how far the people were really overcome with delusions; and as they had a notion of the approach of a visitation, all their predictions ran upon a most dreadful plague, which should lay the whole city, and even the kingdom, waste, and should destroy almost all the nation, both man and beast.

To this, as I said before, the astrologers added stories of the conjunctions of planets in a malignant manner and with a mischievous influence, one of which conjunctions was to happen, and did happen, in October, and the other in November; and they filled the people's heads with predictions on these signs of the heavens, intimating that those conjunctions foretold drought, famine, and pestilence. In the two first of them, however, they were entirely mistaken, for we had no droughty season, but in the beginning of the year a hard frost, which lasted from December almost to March, and after that moderate weather, rather warm than hot, with refreshing winds, and, in short, very seasonable weather, and also several very great rains.

Some endeavours were used to suppress the printing of such books as terrified the people, and to frighten the dispersers of them, some of whom were taken up; but nothing was done in it, as I am informed, the Government being unwilling to exasperate the people, who were, as I may say, all out of their wits already.

Neither can I acquit those ministers that in their sermons rather sank than lifted up the hearts of their hearers. Many of them no doubt did it for the strengthening the resolution of the people, and especially for quickening them to repentance, but it certainly answered not their end, at least not in proportion to the injury it did another way; and indeed, as God Himself through the whole Scriptures rather draws to Him by invitations and calls to turn to Him and live, than drives us by terror and amazement, so I must confess I thought the ministers should have done also, imitating our blessed Lord and Master in this, that His whole Gospel is full of declarations from heaven of God's mercy, and His readiness to receive penitents and forgive them, complaining, 'Ye will not come unto Me that ye may have life', and that therefore His Gospel is called the Gospel of Peace and the Gospel of Grace.

But we had some good men, and that of all persuasions and opinions, whose discourses were full of terror, who spoke nothing but dismal things; and as they brought the people together with a kind of horror, sent them away in tears, prophesying nothing but evil tidings, terrifying the people with the apprehensions of being utterly destroyed, not guiding them, at least not enough, to cry to heaven for mercy.

It was, indeed, a time of very unhappy breaches among us in matters of religion. Innumerable sects and divisions and separate opinions prevailed among the people. The Church of England was restored, indeed, with the restoration of the monarchy, about four years before; but the ministers and preachers of the Presbyterians and Independents, and of all the other sorts of professions, had begun to gather separate societies and erect altar against altar, and all those had their meetings for worship apart, as they have now, but not so many then, the Dissenters being not thoroughly formed into a body as they are since; and those congregations which were thus gathered together were yet but few. And even those that were, the Government did not allow, but endeavoured to suppress them and shut up their meetings.

But the visitation reconciled them again, at least for a time, and many of the best and most valuable ministers and preachers of the Dissenters were suffered to go into the churches where the incumbents were fled away, as many were, not being able to stand it; and the people flocked without distinction to hear them preach, not much inquiring who or what opinion they were of. But after the sickness was over, that spirit of charity abated; and every church being again supplied with their own ministers, or others presented where the minister was dead, things returned to their old channel again.

One mischief always introduces another. These terrors and apprehensions of the people led them into a thousand weak, foolish, and wicked things, which they wanted not a sort of people really wicked to encourage them to: and this was running about to fortune-tellers, cunning-men, and astrologers to know their fortune, or, as it is vulgarly expressed, to have their fortunes told them, their nativities calculated, and the like; and this folly presently made the town swarm with a wicked generation of pretenders to magic, to the black art, as they called it, and I know not what; nay, to a thousand worse dealings with the devil than they were really guilty of. And this trade grew so open and so generally practised that it became common to have signs and inscriptions set up at doors: 'Here lives a fortune-teller', 'Here lives an astrologer', 'Here you may have your nativity calculated', and the like; and Friar Bacon's brazen-head, which was the usual sign of these people's dwellings, was to be seen almost in every street, or else the sign of Mother Shipton, or of Merlin's head, and the like.

With what blind, absurd, and ridiculous stuff these oracles of the devil pleased and satisfied the people I really know not, but certain it is that innumerable attendants crowded about their doors every day. And if but a grave fellow in a velvet jacket, a band, and a black coat, which was the habit those quack-conjurers generally went in, was but seen in the streets the people would follow them in crowds, and ask them questions as they went along.

I need not mention what a horrid delusion this was, or what it tended to; but there was no remedy for it till the plague itself put an end to it all—and, I suppose, cleared the town of most of those calculators themselves. One mischief was, that if the poor people asked these mock astrologers whether there would be a plague or no, they all agreed in general to answer 'Yes', for that kept up their trade. And had the people not been kept in a fright about that, the wizards would presently have been rendered useless, and their craft had been at an end. But they always talked to them of such-and-such influences of the stars, of the conjunctions of such-and-such planets, which must necessarily bring sickness and distempers, and consequently the plague. And some had the assurance to tell them the plague was begun already, which was too true, though they that said so knew nothing of the matter.

The ministers, to do them justice, and preachers of most sorts that were serious and understanding persons, thundered against these and other wicked practices, and exposed the folly as well as the wickedness of them together, and the most sober and judicious people despised and abhorred them. But it was impossible to make any impression upon the middling people and the working labouring poor. Their fears were predominant over all their passions, and they threw away their money in a most distracted manner upon those whimsies. Maid-servants especially, and men-servants, were the chief of their customers, and their question generally was, after the first demand of 'Will there be a plague?' I say, the next question was, 'Oh, sir I for the

Lord's sake, what will become of me? Will my mistress keep me, or will she turn me off? Will she stay here, or will she go into the country? And if she goes into the country, will she take me with her, or leave me here to be starved and undone?' And the like of menservants.

The truth is, the case of poor servants was very dismal, as I shall have occasion to mention again by-and-by, for it was apparent a prodigious number of them would be turned away, and it was so. And of them abundance perished, and particularly of those that these false prophets had flattered with hopes that they should be continued in their services, and carried with their masters and mistresses into the country; and had not public charity provided for these poor creatures, whose number was exceeding great and in all cases of this nature must be so, they would have been in the worst condition of any people in the city.

These things agitated the minds of the common people for many months, while the first apprehensions were upon them, and while the plague was not, as I may say, yet broken out. But I must also not forget that the more serious part of the inhabitants behaved after another manner. The Government encouraged their devotion, and appointed public prayers and days of fasting and humiliation, to make public confession of sin and implore the mercy of God to avert the dreadful judgement which hung over their heads; and it is not to he expressed with what alacrity the people of all persuasions embraced the occasion; how they flocked to the churches and meetings, and they were all so thronged that there was often no coming near, no, not to the very doors of the largest churches. Also there were daily prayers appointed morning and evening at several churches, and days of private praying at other places; at all which the people attended, I say, with an uncommon devotion. Several private families also, as well of one opinion as of another, kept family fasts, to which they admitted their near relations only. So that, in a word, those people who were really serious and religious applied themselves in a truly Christian manner to the proper work of repentance and humiliation, as a Christian people ought to do.

Again, the public showed that they would bear their share in these things; the very Court, which was then gay and luxurious, put on a face of just concern for the public danger. All the plays and interludes which, after the manner of the French Court, had been set up, and began to increase among us, were forbid to act; the gaming-tables, public dancing-rooms, and music-houses, which multiplied and began to debauch the manners of the people, were shut up and suppressed; and the jack-puddings, merry-andrews.puppet-shows, rope-dancers, and such-like doings, which had bewitched the poor common people, shut up their shops, finding indeed no trade; for the minds of the people were agitated with other things, and a kind of sadness and horror at these things sat upon the countenances even of the common people. Death was before their eyes, and everybody began to think of their graves, not of mirth and diversions.

But even those wholesome reflections—which, rightly managed, would have most happily led the people to fall upon their knees, make confession of their sins, and look up to their merciful Saviour for pardon, imploring His compassion on them in such a time of their distress, by which we might have been as a second Nineveh—had a quite contrary extreme in the common people, who, ignorant and stupid in their reflections as they were brutishly wicked and thoughtless before, were now led by their fright to extremes of folly; and, as I have said before, that they ran to conjurers and witches, and all sorts of deceivers, to know what should become of them (who fed their fears, and kept them always alarmed and awake on purpose to delude them and pick their pockets), so they were as mad upon their running after quacks and mountebanks, and every practising old woman, for medicines and remedies; storing themselves with such multitudes of pills, potions, and preservatives, as they were called, that they not only spent their money but even poisoned themselves beforehand for fear of the poison of the infection; and prepared their bodies for the plague, instead of preserving them against it. On the other hand it is incredible and scarce to be imagined, how the posts of houses and corners of streets were plastered over with doctors' bills and papers of ignorant fellows, quacking and tampering in physic, and inviting the people to come to them for remedies, which was generally set off with such flourishes as these, viz.: 'Infallible preventive pills against the plague.' 'Neverfailing preservatives against the infection.' 'Sovereign cordials against the corruption of the air.' 'Exact regulations for the conduct of the body in case of an infection.' 'Anti- pestilential pills.' 'Incomparable drink against the plague, never found out before.' 'An universal remedy for the plague.' 'The only time plague water.' 'The royal antidote against all kinds of infection';—and such a number more that I cannot reckon up; and if I could, would fill a book of themselves to set them down.

Others set up bills to summon people to their lodgings for directions and advice in the case of infection. These had specious titles also, such as these: 'An eminent High Dutch physician, newly come over from Holland, where he resided during all the time of the great plague last year in Amsterdam, and cured multitudes of people that actually had the plague upon them.'

'An Italian gentlewoman just arrived from Naples, having a choice secret to prevent infection, which she found out by her great experience, and did wonderful cures with it in the late plague there, wherein there died 20,000 in one day.'

'An ancient gentlewoman, having practised with great success in the late plague in this city, anno 1636, gives her advice only to the female sex. To be spoken with,' &c.

'An experienced physician, who has long studied the doctrine of antidotes against all sorts of poison and infection, has, after forty years' practice, arrived to such skill as may, with God's blessing, direct persons how to prevent their being touched by any contagious distemper whatsoever. He directs the poor gratis.'

I take notice of these by way of specimen. I could give you two or three dozen of the like and yet have abundance left behind. 'Tis sufficient from these to apprise any one of the humour of those times, and how a set of thieves and pickpockets not only robbed and cheated the poor people of their money, but poisoned their bodies with odious and fatal preparations; some with mercury, and some with other things as bad, perfectly remote from the thing pretended to, and rather hurtful than serviceable to the body in case an infection followed.

I cannot omit a subtility of one of those quack operators, with which he gulled the poor people to crowd about him, but did nothing for them without money. He had, it seems, added to his bills, which he gave about the streets, this advertisement in capital letters, viz., 'He gives advice to the poor for nothing.'

Abundance of poor people came to him accordingly, to whom he made a great many fine speeches, examined them of the state of their health and of the constitution of their bodies, and told them many good things for them to do, which were of no great moment. But the issue and conclusion of all was, that he had a preparation which if they took such a quantity of every morning, he would pawn his life they should never have the plague; no, though they lived in the house with people that were infected. This made the people all resolve to have it; but then the price of that was so much, I think 'twas half-a-crown. 'But, sir,' says one poor woman, 'I am a poor almswoman and am kept by the parish, and your bills say you give the poor your help for nothing.' 'Ay, good woman,' says the doctor, 'so I do, as I published there. I give my advice to the poor for nothing, but not my physic.' 'Alas, sir!' says she, 'that is a snare laid for the poor, then; for you give them advice for nothing; that is to say, you advise them gratis, to buy your physic for their money; so does every shop-keeper with his wares.' Here the woman began to give him ill words, and stood at his door all that day, telling her tale to all the people that came, till the doctor finding she turned away his customers, was obliged to call her upstairs again, and give her his box of physic for nothing, which perhaps, too, was good for nothing when she had it.

But to return to the people, whose confusions fitted them to be imposed upon by all sorts of pretenders and by every mountebank. There is no doubt but these quacking sort of fellows raised great gains out of the miserable people, for we daily found the crowds that ran after them were infinitely greater, and their doors were more thronged than those of Dr Brooks, Dr Upton, Dr Hodges, Dr Berwick, or any, though the most famous men of the time. I And I was told that some of them got five pounds a day by their physic.

But there was still another madness beyond all this, which may serve to give an idea of the distracted humour of the poor people at that time: and this was their following a worse sort of deceivers than any of these; for these petty thieves only deluded them to pick their pockets and get their money, in which their wickedness, whatever it was, lay chiefly on the side of the deceivers, not upon the deceived. But in this part I am going to mention, it lay chiefly in the people deceived, or equally in both; and this was in wearing charms, philtres, exorcisms, amulets, and I know not what preparations, to fortify the body with them against the plague; as if the plague was not the hand of God, but a kind of possession of an evil spirit, and that it was to be kept off with crossings, signs of the zodiac, papers tied up with so many knots, and certain words or figures written on them, as particularly the word Abracadabra, formed in triangle or pyramid, thus:—

ABRACADABRA
ABRACADABR Others had the Jesuits'
ABRACADAB mark in a cross:
ABRACADAI H
ABRACAD S.
ABRACA
ABRAC Others nothing but this
ABRA mark, thus:
ABR
AB **
A{*}
* *

I might spend a great deal of time in my exclamations against the follies, and indeed the wickedness, of those things, in a time of such danger, in a matter of such consequences as this, of a national infection. But my memorandums of these things relate rather to take notice only of the fact, and mention only that it was so. How the poor people found the insufficiency of those things, and how many of them were afterwards carried away in the dead-carts and thrown into the common graves of every parish with these hellish charms and trumpery hanging about their necks, remains to be spoken of as we go along.

All this was the effect of the hurry the people were in, after the first notion of the plague being at hand was among them, and which may be said to be from about Michaelmas 1664, but more particularly after the two men died in St Giles's in the beginning of December; and again, after another alarm in February. For when the plague evidently spread itself, they soon began to see the folly of trusting to those unperforming creatures who had gulled them of their money; and then their fears worked another way, namely, to amazement and stupidity, not knowing what course to take or what to do either to help or relieve themselves. But they ran about from one neighbour's house to another, and even in the streets from one door to another, with repeated cries of, 'Lord, have mercy upon us! What shall we do?'

Indeed, the poor people were to be pitied in one particular thing in which they had little or no relief, and which I desire to mention with a serious awe and reflection, which perhaps every one that reads this may not relish; namely, that whereas death now began not, as we may say, to hover over every one's head only, but to look into their houses and chambers and stare in their faces. Though there might be some stupidity and dullness of the mind (and there was so, a great deal), yet there was a great deal of just alarm sounded into the very inmost soul, if I may so say, of others. Many consciences were awakened; many hard hearts melted into tears; many a penitent confession was made of crimes long concealed. It would wound the soul of any Christian to have heard the dying groans of many a despairing creature, and none durst come near to comfort them. Many a robbery, many a murder, was then confessed aloud, and nobody surviving to record the accounts of it. People might be heard, even into the streets as we passed along, calling upon God for mercy through Jesus Christ, and saying, 'I have been a thief, 'I have been an adulterer', 'I have been a murderer', and the like, and none durst stop to make the least inquiry into such things or to administer comfort to the poor creatures that in the anguish both of soul and body thus cried out. Some of the ministers did visit the sick at first and for a little while, but it was not to be done. It would have been present death to have gone into some houses. The very buriers of the dead, who were the hardenedest creatures in town, were sometimes beaten back and so terrified that they durst not go into houses where the whole families were swept away together, and where the circumstances were more particularly horrible, as some were; but this was, indeed, at the first heat of the distemper.

Time inured them to it all, and they ventured everywhere afterwards without hesitation, as I shall have occasion to mention at large hereafter.

I am supposing now the plague to be begun, as I have said, and that the magistrates began to take the condition of the people into their serious consideration. What they did as to the regulation of the inhabitants and of infected families, I shall speak to by itself; but as to the affair of health, it is proper to mention it here that, having seen the foolish humour of the people in running after quacks and mountebanks, wizards and fortune-tellers, which they did as above, even to madness, the Lord Mayor, a very sober and religious gentleman, appointed physicians and surgeons for relief of the poor—I mean the diseased poor and in particular ordered the College of Physicians to publish directions for cheap remedies for the poor, in all the circumstances of the distemper. This, indeed, was one of the most charitable and judicious things that could be done at that time, for this drove the people from haunting the doors of every disperser of bills, and from taking down blindly and without consideration poison for physic and death instead of life.

This direction of the physicians was done by a consultation of the whole College; and, as it was particularly calculated for the use of the poor and for cheap medicines, it was made public, so that everybody might see it, and copies were given gratis to all that desired it. But as it is public, and to be seen on all occasions, I need not give the reader of this the trouble of it.

I shall not be supposed to lessen the authority or capacity of the physicians when I say that the violence of the distemper, when it came to its extremity, was like the fire the next year. The fire, which consumed what the plague could not touch, defied all the application of remedies; the fire-engines were broken, the buckets thrown away, and the power of man was baffled and brought to an end. So the Plague defied all medicines; the very physicians were seized with it, with their preservatives in their mouths; and men went about prescribing to others and telling them what to do till the tokens were upon them, and they dropped down dead, destroyed by that very enemy they directed others to oppose. This was the case of several physicians, even some of them the most eminent, and of several of the most skillful surgeons. Abundance of quacks too died, who had the folly to trust to their own medicines, which they must needs be conscious to themselves were good for nothing, and who rather ought, like other sorts of thieves, to have run away, sensible of their guilt, from the justice that they could not but expect should punish them as they knew they had deserved.

Not that it is any derogation from the labour or application of the physicians to say they fell in the common calamity; nor is it so intended by me; it rather is to their praise that they ventured their lives so far as even to lose them in the service of mankind. They endeavoured to do good, and to save the lives of others. But we were not to expect that the physicians could stop God's judgements, or prevent a distemper eminently armed from heaven from executing the errand it was sent about.

Doubtless, the physicians assisted many by their skill, and by their prudence and applications, to the saving of their lives and restoring their health. But it is not lessening their character or their skill, to say they could not cure those that had the tokens upon them, or those who were mortally infected before the physicians were sent for, as was frequently the case.

Part 2

It remains to mention now what public measures were taken by the magistrates for the general safety, and to prevent the spreading of the distemper, when it first broke out. I shall have frequent occasion to speak of the prudence of the magistrates, their charity, their vigilance for the poor, and for preserving good order, furnishing provisions, and the like, when the plague was increased, as it afterwards was. But I am now upon the order and regulations they published for the government of infected families.

I mentioned above shutting of houses up; and it is needful to say something particularly to that, for this part of the history of the plague is very melancholy, but the most grievous story must be told.

About June the Lord Mayor of London and the Court of Aldermen, as I have said, began more particularly to concern themselves for the regulation of the city.

The justices of Peace for Middlesex, by direction of the Secretary of State, had begun to shut up houses in the parishes of St Giles-in-the-Fields, St Martin, St Clement Danes, &c., and it was with good success; for in several streets where the plague broke out, upon strict guarding the houses that were infected, and taking care to bury those that died immediately after they were known to be dead, the plague ceased in those streets. It was also observed that the plague decreased sooner in those parishes after they had been visited to the full than it did in the parishes of Bishopsgate, Shoreditch, Aldgate, Whitechappel, Stepney, and others; the early care taken in that manner being a great means to the putting a check to it.

This shutting up of houses was a method first taken, as I understand, in the plague which happened in 1603, at the coming of King James the First to the crown; and the power of shutting people up in their own houses was granted by Act of Parliament, entitled, 'An Act for the charitable Relief and Ordering of Persons infected with the Plague'; on which Act of Parliament the Lord Mayor and aldermen of the city of London founded the order they made at this time, and which took place the 1st of July 1665, when the numbers infected within the city were but few, the last bill for the ninety- two parishes being but four; and some houses having been shut up in the city, and some people being removed to the pest-house beyond Bunhill Fields, in the way to Ishngton—I say, by these means, when there died near one thousand a week in the whole, the number in the city was but twenty-eight, and the city was preserved more healthy in proportion than any other place all the time of the infection.

These orders of my Lord Mayor's were published, as I have said, the latter end of June, and took place from the 1st of July, and were as follows, viz.:—

Orders Conceived and Published by the Lord Mayor and Aldermen of the City of London Concerning the Infection of the Plague, 1665.

'WHEREAS in the reign of our late Sovereign King James, of happy memory, an Act was made for the charitable relief and ordering of persons infected with the plague, whereby authority was given to justices of the peace, mayors, bailiffs, and other head-officers to appoint within their several limits examiners, searchers, watchmen, keepers, and buriers for the persons and places infected, and to minister unto them oaths for the performance of their offices. And the same statute did also authorise the giving of other directions, as unto them for the present necessity should seem good in their directions. It is now, upon special consideration, thought very expedient for preventing and avoiding of infection of sickness (if it shall so please Almighty

God) that these officers following be appointed, and these orders hereafter duly observed.

Examiners to be appointed in every Parish.

'First, it is thought requisite, and so ordered, that in every parish there be one, two, or more persons of good sort and credit chosen and appointed by the alderman, his deputy, and common council of every ward, by the name of examiners, to continue in that office the space of two months at least. And if any fit person so appointed shall refuse to undertake the same, the said parties so refusing to be committed to prison until they shall confonn themselves accordingly.

The Examiner's Office.

'That these examiners he sworn by the aldermen to inquire and learn from time to time what houses in every parish be visited, and what persons be sick, and of what diseases, as near as they can inform themselves; and upon doubt in that case, to command restraint of access until it appear what the disease shall prove. And if they find any person sick of the infection, to give order to the constable that the house be shut up; and if the constable shall be found remiss or negligent, to give present notice thereof to the alderman of the ward.

Watchmen.

'That to every infected house there be appointed two watchmen, one for every day, and the other for the night; and that these watchmen have a special care that no person go in or out of such infected houses whereof they have the charge, upon pain of severe punishment. And the said watchmen to do such further offices as the sick house shall need and require: and if the watchman be sent upon any business, to lock up the house and take the key with him; and the watchman by day to attend until ten of the clock at night, and the watchman by night until six in the morning.

Searchers.

'That there be a special care to appoint women searchers in every parish, such as are of honest reputation, and of the best sort as can be got in this kind; and these to be sworn to make due search and true report to the utmost of their knowledge whether the persons whose bodies they are appointed to search do die of the infection, or of what other diseases, as near as they can. And that the physicians who shall be appointed for cure and prevention of the infection do call before them the said searchers who are, or shall be, appointed for the several parishes under their respective cares, to the end they may consider whether they are fitly qualified for that employment, and charge them from time to time as they shall see cause, if they appear defective in their duties.

'That no searcher during this time of visitation be permitted to use any public work or employment, or keep any shop or stall, or be employed as a laundress, or in any other common employment whatsoever.

Chirurgeons.

'For better assistance of the searchers, forasmuch as there hath been heretofore great abuse in misreporting the disease, to the further spreading of the infection, it is therefore ordered that there be chosen and appointed able and discreet chirurgeons, besides those that do already belong to the pest-house, amongst whom the city and Liberties to be quartered as the places lie most apt and convenient; and every of these to have one quarter for his limit; and the said chirurgeons in every of their limits to join with the searchers for the view of the body, to the end there may be a true report made of the disease.

'And further, that the said chirurgeons shall visit and search such-like persons as shall either send for them or be named and directed unto them by the examiners of every parish, and inform themselves of the disease of the said parties.

'And forasmuch as the said chirurgeons are to be sequestered from all other cures, and kept only to this disease of the infection, it is ordered that every of the said chirurgeons shall have twelve-pence a body searched by them, to be paid out of the goods of the party searched, if he be able, or otherwise by the parish.

Nurse-keepers.

'If any nurse-keeper shall remove herself out of any infected house before twenty-eight days after the decease of any person dying of the infection, the house to which the said nurse-keeper doth so remove herself shall be shut up until the said twenty-eight days be expired.'

Orders Concerning Infected Houses and Persons Sick of the Plague.

Notice to be given of the Sickness.

'The master of every house, as soon as any one in his house complaineth, either of blotch or purple, or swelling in any part of his body, or falleth otherwise dangerously sick, without apparent cause of some other disease, shall give knowledge thereof to the examiner of health within two hours after the said sign shall appear.

Sequestration of the Sick.

'As soon as any man shall be found by this examiner, chirurgeon, or searcher to be sick of the plague, he shall the same night be sequestered in the same house; and in case he be so sequestered, then though he afterwards die not, the house wherein he sickened should be shut up for a month, after the use of the due preservatives taken by the rest.

Airing the Stuff.

'For sequestration of the goods and stuff of the infection, their bedding and apparel and hangings of chambers must be well aired with fire and such perfumes as are requisite within the infected house before they be taken again to use. This to be done by the appointment of an examiner.

Shutting up of the House.

'If any person shall have visited any man known to be infected of the plague, or entered willingly into any known infected house, being not allowed, the house wherein he inhabiteth shall be shut up for certain days by the examiner's direction.

None to be removed out of infected Houses, but, &C.

'Item, that none be removed out of the house where he falleth sick of the infection into any other house in the city (except it be to the pest-house or a tent, or unto some such house which the owner of the said visited house holdeth in his own hands and occupieth by his own servants); and so as security be given to the parish whither such remove is made, that the attendance and charge about the said visited persons shall be observed and charged in all the particularities before expressed, without any cost of that parish to which any such remove shall happen to be made, and this remove to be done by night. And it shall be lawful to any person that hath two houses to remove either his sound or his infected people to his spare house at his choice, so as, if he send away first his sound, he not after send thither his sick, nor again unto the sick the sound; and that the same which he sendeth be for one week at the least shut up and secluded from company, for fear of some infection at the first not appearing.

Burial of the Dead.

'That the burial of the dead by this visitation be at most convenient hours, always either before sun-rising or after sun-setting, with the privity of the churchwardens or constable, and not otherwise; and that no neighbours nor friends be suffered to accompany the corpse to church, or to enter the house visited, upon pain of having his house shut up or be imprisoned.

'And that no corpse dying of infection shall be buried, or remain in any church in time of common prayer, sermon, or lecture. And that no children be suffered at time of burial of any corpse in any church, churchyard, or burying-place to come near the corpse, coffin, or grave. And that all the graves shall be at least six feet deep.

'And further, all public assemblies at other burials are to be forebome during the continuance of this visitation.

No infected Stuff to be uttered.

'That no clothes, stuff, bedding, or garments be suffered to be carried or conveyed out of any infected houses, and that the criers and carriers abroad of bedding or old apparel to be sold or pawned be utterly prohibited and restrained, and no brokers of bedding or old apparel be permitted to make any outward show, or hang forth on their stalls, shop-boards, or windows, towards any street, lane, common way, or passage, any old bedding or apparel to be sold, upon pain of imprisonment. And if any broker or other person shall buy any bedding, apparel, or other stuff out of any infected house within two months after the infection hath been there, his house shall be shut up as infected, and so shall continue shut up twenty days at the least.

No Person to be conveyed out of any infected House.

'If any person visited do fortune, by negligent looking unto, or by any other means, to come or be conveyed from a place infected to any other place, the parish from whence such party hath come or been conveyed, upon notice thereof given, shall at their charge cause the said party so visited and escaped to be carried and brought back again by night, and the parties in this case offending to be punished at the direction of the alderman of the ward, and the house of the receiver of such visited person to be shut up for twenty days.

Every visited House to be marked.

That every house visited be marked with a red cross of a foot long in the middle of the door, evident to be seen, and with these usual printed words, that is to say, "Lord, have mercy upon us," to be set close over the same cross, there to continue until lawful opening of the same house.

Every visited House to be watched.

That the constables see every house shut up, and to be attended with watchmen, which may keep them in, and minister necessaries unto them at their own charges, if they be able, or at the common charge, if they are unable; the shutting up to be for the space of four weeks after all be whole.

'That precise order to be taken that the searchers, chirurgeons, keepers, and buriers are not to pass the streets without holding a red rod or wand of three feet in length in their hands, open and evident to be seen, and are not to go into any other house than into their own, or into that whereunto they are directed or sent for; but to forbear and abstain from company, especially when they have been lately used in any such business or attendance.

Inmates.

'That where several inmates are in one and the same house, and any person in that house happens to be infected, no other person or family of such house shall be suffered to remove him or themselves without a certificate from the examiners of health of that parish; or in default thereof, the house whither he or they so remove shall be shut up as in case of visitation.

Hackney-Coaches.

'That care be taken of hackney-coachmen, that they may not (as some of them have been observed to do after carrying of infected persons to the pest-house and other places) be admitted to common use till their coaches be well aired, and have stood unemployed by the space of five or six days after such service.'

Orders for Cleansing and Keeping of the Streets Sweet.

The Streets to be kept Clean.

'First, it is thought necessary, and so ordered, that every householder do cause the street to be daily prepared before his door, and so to keep it clean swept all the week long.

That Rakers take it from out the Houses.

That the sweeping and filth of houses be daily carried away by the rakers, and that the raker shall give notice of his coming by the blowing of a horn, as hitherto hath been done.

Laystalls to be made far off from the City.

That the laystalls be removed as far as may be out of the city and common passages, and that no nightman or other be suffered to empty a vault into any garden near about the city.

Care to be had of unwholesome Fish or Flesh, and of musty Corn.

'That special care be taken that no stinking fish, or unwholesome flesh, or musty corn, or other corrupt fruits of what sort soever, be suffered to be sold about the city, or any part of the same.

'That the brewers and tippling-houses he looked unto for musty and unwholesome casks.

'That no hogs, dogs, or cats, or tame pigeons, or conies, be suffered to be kept within any part of the city, or any swine to be or stray in the streets or lanes, but that such swine be impounded by the beadle or any other officer, and the owner punished according to Act of Common Council, and that the dogs be killed by the dog-killers appointed for that purpose.'

Orders Concerning Loose Persons and Idle Assemblies.

Beggars.
'Forasmuch as nothing is more complained of than the multitude of rogues and wandering beggars that swarm in every place about the city, being a great cause of the spreading of the infection, and will not be avoided, notwithstanding any orders that have been given to the contrary: It is therefore now ordered, that such constables, and others whom this matter may any way concern, take special care that no wandering beggars be suffered in the streets of this city in any fashion or manner whatsoever, upon the penalty provided by the law, to be duly and severely executed upon them.

Plays.
'That all plays, bear-baitings, games, singing of ballads, buckler-play, or such-like causes of assemblies of people be utterly prohibited, and the parties offending severely punished by every alderman in his ward.

Feasting prohibited.

'That all public feasting, and particularly by the companies of this city, and dinners at taverns, ale-houses, and other places of common entertainment be forborne till further order and allowance; and that the money thereby spared be preserved and employed for the benefit and relief of the poor visited with the infection.

Tippling-houses.

'That disorderly tippling in taverns, ale-houses, coffee-houses, and cellars be severely looked unto, as the common sin of this time and greatest occasion of dispersing the plague. And that no company or person be suffered to remain or come into any tavern, ale-house, or coffee-house to drink after nine of the clock in the evening, according to the ancient law and custom of this city, upon the penalties ordained in that behalf.

'And for the better execution of these orders, and such other rules and directions as, upon further consideration, shall be found needful: It is ordered and enjoined that the aldermen, deputies, and common councilmen shall meet together weekly, once, twice, thrice or oftener (as cause shall require), at some one general place accustomed in their respective wards (being clear from infection of the plague), to consult how the said orders may be duly put in execution; not intending that any dwelling in or near places infected shall come to the said meeting while their coming may be doubtful. And the said aldermen, and deputies, and common councilmen in their several wards may put in execution any other good orders that by them at their said meetings shall be conceived and devised for preservation of his Majesty's subjects from the infection.

'SIR JOHN LAWRENCE, Lord Mayor.
SIR GEORGE WATERMAN
SIR CHARLES DoE, Sheriffs.'

I need not say that these orders extended only to such places as were within the Lord Mayor's jurisdiction, so it is requisite to observe that the justices of Peace within those parishes and places as were called the Hamlets and out- parts took the same method. As I remember, the orders for shutting up of houses did not take Place so soon on our side, because, as I said before, the plague did not reach to these eastern parts of the town at least, nor begin to be very violent, till the beginning of August. For example, the whole bill from the 11th to the 18th of July was 1761, yet there died but 71 of the plague in all those parishes we call the Tower Hamlets, and they were as follows:—

The next week And to the 1st was thus: of Aug. thus:
Aldgate 14 34 65
Stepney 33 58 76
Whitechappel 21 48 79
St Katherine, Tower 2 4 4
Trinity, Minories 1 1 4

---

71 145 228

It was indeed coming on again, for the burials that same week were in the next adjoining parishes thus:—

The next week
prodigiously To the 1st of
increased, as: Aug. thus:
St Leonard's, Shoreditch 64 84 110
St Botolph's, Bishopsgate 65 105 116
St Giles's, Cripplegate 213 421 554

---

342 610 780

This shutting up of houses was at first counted a very cruel and unchristian method, and the poor people so confined made bitter lamentations. Complaints of the severity of it were also daily brought to my Lord Mayor, of houses causelessly (and some maliciously) shut up. I cannot say; but upon inquiry many that complained so loudly were found in a condition to be continued; and others again, inspection being made upon the sick person, and the sickness not appearing infectious, or if uncertain, yet on his being content to be carried to the pest-house, were released.

It is true that the locking up the doors of people's houses, and setting a watchman there night and day to prevent their stirring out or any coming to them, when perhaps the sound people in the family might have escaped if they had been removed from the sick, looked very hard and cruel; and many people perished in these miserable confinements which, 'tis reasonable to beheve, would not have been distempered if they had had liberty, though the plague was in the house; at which the people were very clamorous and uneasy at first, and several violences were committed and injuries offered to the men who were set to watch the houses so shut up; also several people broke out by force in many places, as I shall observe by-and-by. But it was a public good that justified the private mischief, and there was no obtaining the least mitigation by any application to magistrates or government at that time, at least not that I heard of. This put the people upon all manner of stratagem in order, if possible, to get out; and it would fill a little volume to set down the arts used by the people of such houses to shut the eyes of the watchmen who were employed, to deceive them, and to escape or break out from them, in which frequent scuffles and some mischief happened; of which by itself.

As I went along Houndsditch one morning about eight o'clock there was a great noise. It is true, indeed, there was not much crowd, because people were not very free to gather together, or to stay long together when they were there; nor did I stay long there. But the outcry was loud enough to prompt my curiosity, and I called to one that looked out of a window, and asked what was the matter.

A watchman, it seems, had been employed to keep his post at the door of a house which was infected, or said to be infected, and was shut up. He had been there all night for two nights together, as he told his story, and the day-watchman had been there one day, and was now come to relieve him.

All this while no noise had been heard in the house, no light had been seen; they called for nothing, sent him of no errands, which used to be the chief business of the watchmen; neither had they given him any disturbance, as he said, from the Monday afternoon, when he heard great crying and screaming in the house, which, as he supposed, was occasioned by some of the family dying just at that time. It seems, the night before, the dead-cart, as it was called, had been stopped there, and a servant-maid had been brought down to the door dead, and the buriers or bearers, as they were called, put her into the cart, wrapt only in a green rug, and carried her away.

The watchman had knocked at the door, it seems, when he heard that noise and crying, as above, and nobody answered a great while; but at last one looked out and said with an angry, quick tone, and yet a kind of crying voice, or a voice of one that was crying, 'What d'ye want, that ye make such a knocking?' He answered, 'I am the watchman! How do you do? What is the matter?' The person answered, 'What is that to you? Stop the dead-cart.' This, it seems, was about one o'clock. Soon after, as the fellow said, he stopped the dead-cart, and then knocked again, but nobody answered. He continued knocking, and the bellman called out several times, 'Bring out your dead'; but nobody answered, till the man that drove the cart, being called to other houses, would stay no longer, and drove away.

The watchman knew not what to make of all this, so he let them alone till the moming-man or day-watchman, as they called him, came to relieve him. Giving him an account of the particulars, they knocked at the door a great while, but nobody answered; and they observed that the window or casement at which the person had looked out who had answered before continued open, being up two pair of stairs.

Upon this the two men, to satisfy their curiosity, got a long ladder, and one of them went up to the window and looked into the room, where he saw a woman lying dead upon the floor in a dismal manner, having no clothes on her but her shift. But though he called aloud, and putting in his long staff, knocked hard on the floor, yet nobody stirred or answered; neither could he hear any noise in the house.

He came down again upon this, and acquainted his fellow, who went up also; and finding it just so, they resolved to acquaint either the Lord Mayor or some other magistrate of it, but did not offer to go in at the window. The magistrate, it seems, upon the information of the two men, ordered the house to be broke open, a constable and other persons being appointed to be present, that nothing might be plundered; and accordingly it was so done, when nobody was found in the house but that young woman, who having been infected and past recovery, the rest had left her to die by herself, and were every one gone, having found some way to delude the watchman, and to get open the door, or get out at some back-door, or over the tops of the houses, so that he knew nothing of it; and as to those cries and shrieks which he heard, it was supposed they were the passionate cries of the family at the bitter parting, which, to be sure, it was to them all, this being the sister to the mistress of the family. The man of the house, his wife, several children, and servants, being all gone and fled, whether sick or sound, that I could never learn; nor, indeed, did I make much inquiry after it.

Many such escapes were made out of infected houses, as particularly when the watchman was sent of some errand; for it was his business to go of any errand that the family sent him of; that is to say, for necessaries, such as food and physic; to fetch physicians, if they would come, or surgeons, or nurses, or to order the dead-cart, and the like; but with this condition, too, that when he went he was to lock up the outer door of the house and take the key away with him. To evade this, and cheat the watchmen, people got two or three keys made to their locks, or they found ways to unscrew the locks such as were screwed on, and so take off the lock, being in the inside of the house, and while they sent away the watchman to the market, to the bakehouse, or for one trifle or another, open the door and go out as often as they pleased. But this being found out, the officers afterwards had orders to padlock up the doors on the outside, and place bolts on them as they thought fit.

At another house, as I was informed, in the street next within Aldgate, a whole family was shut up and locked in because the maid-servant was taken sick. The master of the house had complained by his friends to the next alderman and to the Lord Mayor, and had consented to have the maid carried to the pest-house, but was refused; so the door was marked with a red cross, a padlock on the outside, as above, and a watchman set to keep the door, according to public order.

After the master of the house found there was no remedy, but that he, his wife, and his children were to be locked up with this poor distempered servant, he called to the watchman, and told him he must go then and fetch a nurse for them to attend this poor girl, for that it would be certain death to them all to oblige them to nurse her; and told him plainly that if he would not do this, the maid must perish either of the distemper or be starved for want of food, for he was resolved none of his family should go near her; and she lay in the garret four storey high, where she could not cry out, or call to anybody for help.

The watchman consented to that, and went and fetched a nurse, as he was appointed, and brought her to them the same evening. During this interval the master of the house took his opportunity to break a large hole through his shop into a bulk or stall, where formerly a cobbler had sat, before or under his shop-window; but the tenant, as may be supposed at such a dismal time as that, was dead or removed, and so he had the key in his own keeping. Having made his way into this stall, which he could not have done if the man had been at the door, the noise he was obliged to make being such as would have alarmed the watchman; I say, having made his way into this stall, he sat still till the watchman returned with the nurse, and all the next day also. But the night following, having contrived to send the watchman of another trifling errand, which, as I take it, was to an apothecary's for a plaister for the maid, which he was to stay for the making up, or some other such errand that might secure his staying some time; in that time he conveyed himself and all his family out of the house, and left the nurse and the watchman to bury the poor wench—that is, throw her into the cart—and take care of the house.

I could give a great many such stories as these, diverting enough, which in the long course of that dismal year I met with—that is, heard of—and which are very certain to be true, or very near the truth; that is to say, true in the general: for no man could at such a time learn all the particulars. There was likewise violence used with the watchmen, as was reported, in abundance of places; and I beheve that from the beginning of the visitation to the end, there was not less than eighteen or twenty of them killed, or so wounded as to be taken up for dead, which was supposed to be done by the people in the infected houses which were shut up, and where they attempted to come out and were opposed.

Nor, indeed, could less be expected, for here were so many prisons in the town as there were houses shut up; and as the people shut up or imprisoned so were guilty of no crime, only shut up because miserable, it was really the more intolerable to them.

It had also this difference, that every prison, as we may call it, had but one jailer, and as he had the whole house to guard, and that many houses were so situated as that they had several ways out, some more, some less, andsome into several streets, it was impossible for one man so to guard all the passages as to prevent the escape of people made desperate by the fright of their circumstances, by the resentment of their usage, or by the raging of the distemper itself; so that they would talk to the watchman on one side of the house, while the family made their escape at another.

For example, in Coleman Street there are abundance of alleys, as appears still. A house was shut up in that they call White's Alley; and this house had a back-window, not a door, into a court which had a passage into Bell Alley. A watchman was set by the constable at the door of this house, and there he stood, or his comrade, night and day, while the family went all away in the evening out at that window into the court, and left the poor fellows warding and watching for near a fortnight.

Not far from the same place they blew up a watchman with gunpowder, and burned the poor fellow dreadfully; and while he made hideous cries, and nobody would venture to come near to help him, the whole family that were able to stir got out at the windows one storey high, two that were left sick calling out for help. Care was taken to give them nurses to look after them, but the persons fled were never found, till after the plague was abated they returned; but as nothing could be proved, so nothing could be done to them.

It is to be considered, too, that as these were prisons without bars and bolts, which our common prisons are furnished with, so the people let themselves down out of their windows, even in the face of the watchman, bringing swords or pistols in their hands, and threatening the poor wretch to shoot him if he stirred or called for help.

In other cases, some had gardens, and walls or pales, between them and their neighbours, or yards and back-houses; and these, by friendship and entreaties, would get leave to get over those walls or pales, and so go out at their neighbours' doors; or, by giving money to their servants, get them to let them through in the night; so that in short, the shutting up of houses was in no wise to be depended upon. Neither did it answer the end at all, serving more to make the people desperate, and drive them to such extremities as that they would break out at all adventures.

And that which was still worse, those that did thus break out spread the infection farther by their wandering about with the distemper upon them, in their desperate circumstances, than they would otherwise have done; for whoever considers all the particulars in such cases must acknowledge, and we cannot doubt but the severity of those confinements made many people desperate, and made them run out of their houses at aU hazards, and with the plague visibly upon them, not knowing either whither to go or what to do, or, indeed, what they did; and many that did so were driven to dreadful exigencies and extremities, and perished in the streets or fields for mere want, or dropped down by the raging violence of the fever upon them. Others wandered into the country, and went forward any way, as their desperation guided them, not knowing whither they went or would go: till, faint and tired, and not getting any relief, the houses and villages on the road refusing to admit them to lodge whether infected or no, they have perished by the roadside or gotten into bams and died there, none daring to come to them or relieve them, though perhaps not infected, for nobody would believe them.

On the other hand, when the plague at first seized a family that is to say, when any body of the family had gone out and unwarily or otherwise catched the distemper and brought it home—it was certainly known by the family before it was known to the officers, who, as you will see by the order, were appointed to examine into the circumstances of all sick persons when they heard of their being sick.

In this interval, between their being taken sick and the examiners coming, the master of the house had leisure and liberty to remove himself or all his family, if he knew whither to go, and many did so. But the great disaster was that many did thus after they were really infected themselves, and so carried the disease into the houses of those who were so hospitable as to receive them; which, it must be confessed, was very cmel and ungrateful.

And this was in part the reason of the general notion, or scandal rather, which went about of the temper of people infected: namely, that they did not take the least care or make any scmple of infecting others, though I cannot say but there might be some truth in it too, but not so general as was reported. What natural reason could be given for so wicked a thing at a time when they might conclude themselves just going to appear at the bar of Divine justice I know not. I am very well satisfied that it cannot be reconciled to religion and principle any more than it can be to generosity and Humanity, but I may speak of that again.

I am speaking now of people made desperate by the apprehensions of their being shut up, and their breaking out by stratagem or force, either before or after they were shut up, whose misery was not lessened when they were out, but sadly increased. On the other hand, many that thus got away had retreats to go to and other houses, where they locked themselves up and kept hid till the plague was over; and many families, foreseeing the approach of the distemper, laid up stores of provisions sufficient for their whole families, and shut themselves up, and that so entirely that they were neither seen or heard of till the infection was quite ceased, and then came abroad sound and well. I might recollect several such as these, and give you the particulars of their management; for doubtless it was the most effectual secure step that could be taken for such whose circumstances would not admit them to remove, or who had not retreats abroad proper for the case; for in being thus shut up they were as if they had been a hundred miles off. Nor do I remember that any one of those families miscarried. Among these, several Dutch merchants were particularly remarkable, who kept their houses like little garrisons besieged suffering none to go in or out or come near them, particularly one in a court in Throgmorton Street whose house looked into Draper's Garden.

But I come back to the case of families infected and shut up by the magistrates. The misery of those families is not to be expressed; and it was generally in such houses that we heard the most dismal shrieks and outcries of the poor people, terrified and even frighted to death by the sight of the condition of their dearest relations, and by the terror of being imprisoned as they were.

I remember, and while I am writing this story I think I hear the very sound of it, a certain lady had an only daughter, a young maiden about nineteen years old, and who was possessed of a very considerable fortune. They were only lodgers in the house where they were. The young woman, her mother, and the maid had been abroad on some occasion, I do not remember what, for the house was not shut up; but about two hours after they came home the young lady complained she was not well; in a quarter of an hour more she vomited and had a violent pain in her head. 'Pray God', says her mother, in a terrible fright, 'my child has not the distemper!' The pain in her head increasing, her mother ordered the bed to be warmed, and resolved to put her to bed, and prepared to give her things to sweat, which was the ordinary remedy to be taken when the first apprehensions of the distemper began.

While the bed was airing the mother undressed the young woman, and just as she was laid down in the bed, she, looking upon her body with a candle, immediately discovered the fatal tokens on the inside of her thighs. Her mother, not being able to contain herself, threw down her candle and shrieked out in such a frightful manner that it was enough to place horror upon the stoutest heart in the world; nor was it one scream or one cry, but the fright having seized her spirits, she fainted first, then recovered, then ran all over the house, up the stairs and down the stairs, like one distracted, and indeed really was distracted, and continued screeching and crying out for several hours void of all sense, or at least government of her senses, and, as I was told, never came thoroughly to herself again. As to the young maiden, she was a dead corpse from that moment, for the gangrene which occasions the spots had spread [over] her whole body, and she died in less than two hours. But still the mother continued crying out, not knowing anything more of her child, several hours after she was dead. It is so long ago that I am not certain, but I think the mother never recovered, but died in two or three weeks after.

This was an extraordinary case, and I am therefore the more particular in it, because I came so much to the knowledge of it; but there were innumerable such-like cases, and it was seldom that the weekly bill came in but there were two or three put in, 'frighted'; that is, that may well be called frighted to death. But besides those who were so frighted as to die upon the spot, there were great numbers frighted to other extremes, some frighted out of their senses, some out of their memory, and some out of their understanding. But I return to the shutting up of houses.

As several people, I say, got out of their houses by stratagem after they were shut UP, so others got out by bribing the watchmen, and giving them money to let them go privately out in the night. I must confess I thought it at that time the most innocent corruption or bribery that any man could be guilty of, and therefore could not but pity the poor men, and think it was hard when three of those watchmen were publicly whipped through the streets for suffering people to go out of houses shut up.

But notwithstanding that severity, money prevailed with the poor men, and many families found means to make sallies out, and escape that way after they had been shut up; but these were generally such as had some places to retire to; and though there was no easy passing the roads any whither after the 1st of August, yet there were many ways of retreat, and particularly, as I hinted, some got tents and set them up in the fields, carrying beds or straw to lie on, and provisions to eat, and so lived in them as hermits in a cell, for nobody would venture to come near them; and several stories were told of such, some comical, some tragical, some who lived like wandering pilgrims in the deserts, and escaped by making themselves exiles in such a manner as is scarce to be credited, and who yet enjoyed more liberty than was to be expected in such cases.

I have by me a story of two brothers and their kinsman, who being single men, but that had stayed in the city too long to get away, and indeed not knowing where to go to have any retreat, nor having wherewith to travel far, took a course for their own preservation, which though in itself at first desperate, yet was so natural that it may be wondered that no more did so at that time. They were but of mean condition, and yet not so very poor as that they could not furnish themselves with some little conveniences such as might serve to keep life and soul together; and finding the distemper increasing in a terrible manner, they resolved to shift as well as they could, and to be gone.

One of them had been a soldier in the late wars, and before that in the Low Countries, and having been bred to no particular employment but his arms, and besides being wounded, and not able to work very hard, had for some time been employed at a baker's of sea-biscuit in Wapping.

The brother of this man was a seaman too, but somehow or other had been hurt of one leg, that he could not go to sea, but had worked for his living at a sailmaker's in Wapping, or thereabouts; and being a good husband, had laid up some money, and was the richest of the three.

The third man was a joiner or carpenter by trade, a handy fellow, and he had no wealth but his box or basket of tools, with the help of which he could at any time get his living, such a time as this excepted, wherever he went—and he lived near Shadwell.

They all lived in Stepney parish, which, as I have said, being the last that was infected, or at least violently, they stayed there till they evidently saw the plague was abating at the west part of the town, and coming towards the east, where they lived.

The story of those three men, if the reader will be content to have me give it in their own persons, without taking upon me to either vouch the particulars or answer for any mistakes, I shall give as distinctly as I can, believing the history will be a very good pattern for any poor man to follow, in case the like public desolation should happen here; and if there may be no such occasion, which God of His infinite mercy grant us, still the story may have its uses so many ways as that it will, I hope, never be said that the relating has been unprofitable.

I say all this previous to the history, having yet, for the present, much more to say before I quit my own part.

I went all the first part of the time freely about the streets, though not so freely as to run myself into apparent danger, except when they dug the great pit in the churchyard of our parish of Aldgate. A terrible pit it was, and I could not resist my curiosity to go and see it. As near as I may judge, it was about forty feet in length, and about fifteen or sixteen feet broad, and at the time I first looked at it, about nine feet deep; but it was said they dug it near twenty feet deep afterwards in one part of it, till they could go no deeper for the water; for they had, it seems, dug several large pits before this. For though the plague was long a-coming to our parish, yet, when it did come, there was no parish in or about London where it raged with such violence as in the two parishes of Aldgate and Whitechappel.

I say they had dug several pits in another ground, when the distemper began to spread in our parish, and especially when the dead-carts began to go about, which was not, in our parish, till the beginning of August. Into these pits they had put perhaps fifty or sixty bodies each; then they made larger holes wherein they buried all that the cart brought in a week, which, by the middle to the end of August, came to from 200 to 400 a week; and they could not well dig them larger, because of the order of the magistrates confining them to leave no bodies within six feet of the surface; and the water coming on at about seventeen or eighteen feet, they could not well, I say, put more in one pit. But now, at the beginning of September, the plague raging in a dreadful manner, and the number of burials in our parish increasing to more than was ever buried in any parish about London of no larger extent, they ordered this dreadful gulf to be dug—for such it was, rather than a pit.

They had supposed this pit would have supplied them for a month or more when they dug it, and some blamed the churchwardens for suffering such a frightful thing, telling them they were making preparations to bury the whole parish, and the like; but time made it appear the churchwardens knew the condition of the parish better than they did: for, the pit being finished the 4th of September, I think, they began to bury in it the 6th, and by the 20th, which was just two weeks, they had thrown into it

1114 bodies when they were obliged to fill it up, the bodies being then come to he within six feet of the surface. I doubt not but there may be some ancient persons alive in the parish who can justify the fact of this, and are able to show even in what place of the churchyard the pit lay better than I can. The mark of it also was many years to be seen in the churchyard on the surface, lying in length parallel with the passage which goes by the west wall of the churchyard out of Houndsditch, and turns east again into Whitechappel, coming out near the Three Nuns' Inn.

It was about the 10th of September that my curiosity led, or rather drove, me to go and see this pit again, when there had been near 400 people buried in it; and I was not content to see it in the day-time, as I had done before, for then there would have been nothing to have been seen but the loose earth; for all the bodies that were thrown in were immediately covered with earth by those they called the buriers, which at other times were called bearers; but I resolved to go in the night and see some of them thrown in.

There was a strict order to prevent people coming to those pits, and that was only to prevent infection. But after some time that order was more necessary, for people that were infected and near their end, and delirious also, would run to those pits, wrapt in blankets or rugs, and throw themselves in, and, as they said, bury themselves. I cannot say that the officers suffered any willingly to lie there; but I have heard that in a great pit in Finsbury, in the parish of Cripplegate, it lying open then to the fields, for it was not then walled about, [many] came and threw themselves in, and expired there, before they threw any earth upon them; and that when they came to bury others and found them there, they were quite dead, though not cold.

This may serve a little to describe the dreadful condition of that day, though it is impossible to say anything that is able to give a true idea of it to those who did not see it, other than this, that it was indeed very, very, very dreadful, and such as no tongue can express.

I got admittance into the churchyard by being acquainted with the sexton who attended; who, though he did not refuse me at all, yet earnestly persuaded me not to go, telling me very seriously (for he was a good, religious, and sensible man) that it was indeed their business and duty to venture, and to run aU hazards, and that in it they might hope to be preserved; but that I had no apparent call to it but my own curiosity, which, he said, he believed I would not pretend was sufficient to justify my running that hazard. I told him I had been pressed in my mind to go, and that perhaps it might be an instructing sight, that might not be without its uses. 'Nay,' says the good man, 'if you will venture upon that score, name of God go in; for, depend upon it, 'twill be a sermon to you, it may be, the best that ever you heard in your life. 'Tis a speaking sight,' says he, 'and has a voice with it, and a loud one, to call us all to repentance'; and with that he opened the door and said, 'Go, if you will.'

His discourse had shocked my resolution a little, and I stood wavering for a good while, but just at that interval I saw two links come over from the end of the Minories, and heard the bellman, and then appeared a dead-cart, as they called it, coming over the streets; so I could no longer resist my desire of seeing it, and went in. There was nobody, as I could perceive at first, in the churchyard, or going into it, but the buriers and the fellow that drove the cart, or rather led the horse and cart; but when they came up to the pit they saw a man go to and again, muffled up in a brown Cloak, and making motions with his hands under his cloak, as if he was in great agony, and the buriers immediately gathered about him, supposing he was one of those poor delirious or desperate creatures that used to pretend, as I have said, to bury themselves. He said nothing as he walked about, but two or three times groaned very deeply and loud, and sighed as he would break his heart.

When the buriers came up to him they soon found he was neither a person infected and desperate, as I have observed above, or a person distempered in mind, but one oppressed with a dreadful weight of grief indeed, having his wife and several of his children all in the cart that was just come in with him, and he followed in an agony and excess of sorrow. He mourned heartily, as it was easy to see, but with a kind of masculine grief that could not give itself vent by tears; and calmly defying the buriers to let him alone, said he would only see the bodies thrown in and go away, so they left importuning him. But no sooner was the cart turned round and the bodies shot into the pit promiscuously, which was a surprise to him, for he at least expected they would have been decently laid in, though indeed he was afterwards convinced that was impracticable; I say, no sooner did he see the sight but he cried out aloud, unable to contain himself. I could not hear what he said, but he went backward two or three steps and fell down in a swoon. The buriers ran to him and took him up, and in a little while he came to himself, and they led him away to the Pie Tavern over against the end of Houndsditch, where, it seems, the man was known, and where they took care of him. He looked into the pit again as he went away, but the buriers had covered the bodies so immediately with throwing in earth, that though there was light enough, for there were lanterns, and candles in them, placed all night round the sides of the pit, upon heaps of earth, seven or eight, or perhaps more, yet nothing could be seen.

This was a mournful scene indeed, and affected me almost as much as the rest; but the other was awful and full of terror. The cart had in it sixteen or seventeen bodies; some were wrapt up in linen sheets, some in rags, some little other than naked, or so loose that what covering they had fell from them in the shooting out of the cart, and they fell quite naked among the rest; but the matter was not much to them, or the indecency much to any one else, seeing they were all dead, and were to be huddled together into the common grave of mankind, as we may call it, for here was no difference made, but poor and rich went together; there was no other way of burials, neither was it possible there should, for coffins were not to be had for the prodigious numbers that fell in such a calamity as this.

It was reported by way of scandal upon the buriers, that if any corpse was delivered to them decently wound up, as we called it then, in a winding- sheet tied over the head and feet, which some did, and which was generally of good linen; I say, it was reported that the buriers were so wicked as to strip them in the cart and carry them quite naked to the ground. But as I cannot easily credit anything so vile among Christians, and at a time so filled with terrors as that was, I can only relate it and leave it undetermined.

Innumerable stories also went about of the cruel behaviours and practices of nurses who tended the sick, and of their hastening on the fate of those they tended in their sickness. But I shall say more of this in its place.

I was indeed shocked with this sight; it almost overwhelmed me, and I went away with my heart most afflicted, and full of the afflicting thoughts, such as I cannot describe, just at my going out of the church, and turning up the street towards my own house, I saw another cart with links, and a bellman going before, coming out of Harrow Alley in the Butcher Row, on the other side of the way, and being, as I perceived, very full of dead bodies, it went directly over the street also toward the church. I stood a while, but I had no stomach to go back again to see the same dismal scene over again, so I went directly home, where I could not but consider with thankfulness the risk I had run, believing I had gotten no injury, as indeed I had not.

Here the poor unhappy gentleman's grief came into my head again, and indeed I could not but shed tears in the reflection upon it, perhaps more than he did himself; but his case lay so heavy upon my mind that I could not prevail with myself, but that I must go out again into the street, and go to the Pie Tavern, resolving to inquire what became of him.

It was by this time one o'clock in the morning, and yet the poor gentleman was there. The truth was, the people of the house, knowing him, had entertained him, and kept him there all the night, notwithstanding the danger of being infected by him, though it appeared the man was perfectly sound himself.

It is with regret that I take notice of this tavern. The people were civil, mannerly, and an obhging sort of folks enough, and had till this time kept their house open and their trade going on, though not so very publicly as formerly: but there was a dreadful set of fellows that used their house, and who, in the middle of all this horror, met there every night, behaved with all the revelling and roaring extravagances as is usual for such people to do at other times, and, indeed, to such an offensive degree that the very master and mistress of the house grew first ashamed and then terrified at them.

They sat generally in a room next the street, and as they always kept late hours, so when the dead-cart came across the street-end to go into Houndsditch, which was in view of the tavern windows, they would frequently open the windows as soon as they heard the bell and look out at them; and as they might often hear sad lamentations of people in the streets or at their windows as the carts went along, they would make their impudent mocks and jeers at them, especially if they heard the poor people call upon God to have mercy upon them, as many would do at those times in their ordinary passing along the streets.

These gentlemen, being something disturbed with the clutter of bringing the poor gentleman into the house, as above, were first angry and very high with the master of the house for suffering such a fellow, as they called him, to be brought out of the grave into their house; but being answered that the man was a neighbour, and that he was sound, but overwhelmed with the calamity of his family, and the like, they turned their anger into ridiculing the man and his sorrow for his wife and children, taunted him with want of courage to leap into the great pit and go to heaven, as they jeeringly expressed it, along with them, adding some very profane and even blasphemous expressions.

They were at this vile work when I came back to the house, and, as far as I could see, though the man sat still, mute and disconsolate, and their affronts could not divert his sorrow, yet he was both grieved and offended at their discourse. Upon this I gently reproved them, being well enough acquainted with their characters, and not unknown in person to two of them.

They immediately fell upon me with ill language and oaths, asked me what I did out of my grave at such a time when so many honester men were carried into the churchyard, and why I was not at home saying my prayers against the dead-cart came for me, and the like.

I was indeed astonished at the impudence of the men, though not at all discomposed at their treatment of me. However, I kept my temper. I told them that though I defied them or any man in the world to tax me with any dishonesty, yet I acknowledged that in this terrible judgement of God many better than I were swept away and carried to their grave. But to answer their question directly, the case was, that I was mercifully preserved by that great God whose name they had blasphemed and taken in vain by cursing and swearing in a dreadful manner, and that I believed I was preserved in particular, among other ends of His goodness, that I might reprove them for their audacious boldness in behaving in such a manner and in such an awful time as this was, especially for their jeering and mocking at an honest gentleman and a neighbour (for some of them knew him), who, they saw, was overwhelmed with sorrow for the breaches which it had pleased God to make upon his family.

I cannot call exactly to mind the hellish, abominable raillery which was the return they made to that talk of mine: being provoked, it seems, that I was not at all afraid to be free with them; nor, if I could remember, would I fill my account with any of the words, the horrid oaths, curses, and vile expressions, such as, at that time of the day, even the worst and ordinariest people in the street would not use; for, except such hardened creatures as these, the most wicked wretches that could be found had at that time some terror upon their minds of the hand of that Power which could thus in a moment destroy them.

But that which was the worst in all their devilish language was, that they were not afraid to blaspheme God and talk atheistically, making a jest of my calling the plague the hand of God; mocking, and even laughing, at the word judgement, as if the providence of God had no concern in the inflicting such a desolating stroke; and that the people calling upon God as they saw the carts carrying away the dead bodies was all enthusiastic, absurd, and impertinent.

I made them some reply, such as I thought proper, but which I found was so far from putting a check to their horrid way of speaking that it made them rail the more, so that I confess it filled me with horror and a kind of rage, and I came away, as I told them, lest the hand of that judgement which had visited the whole city should glorify His vengeance upon them, and all that were near them.

They received all reproof with the utmost contempt, and made the greatest mockery that was possible for them to do at me, giving me all the opprobrious, insolent scoffs that they could think of for preaching to them, as they called it, which indeed grieved me, rather than angered me; and I went away, blessing God, however, in my mind that I had not spared them, though they had insulted me so much.

They continued this wretched course three or four days after this, continually mocking and jeering at all that showed themselves religious or serious, or that were any way touched with the sense of the terrible judgement of God upon us; and I was informed they flouted in the same manner at the good people who, notwithstanding the contagion, met at the church, fasted, and prayed to God to remove His hand from them.

I say, they continued this dreadful course three or four days—I think it was no more—when one of them, particularly he who asked the poor gentleman what he did out of his grave, was struck from Heaven with the plague, and died in a most deplorable manner; and, in a word, they were every one of them carried into the great pit which I have mentioned above, before it was quite filled up, which was not above a fortnight or thereabout.

These men were guilty of many extravagances, such as one would think human nature should have trembled at the thoughts of at such a time of general terror as was then upon us, and particularly scoffing and mocking at everything which they happened to see that was religious among the people, especially at their thronging zealously to the place of public worship to implore mercy from Heaven in such a time of distress; and this tavern where they held their dub being within view of the church-door, they had the more particular occasion for their atheistical profane mirth.

But this began to abate a little with them before the accident which I have related happened, for the infection increased so violently at this part of the town now, that people began to be afraid to come to the church; at least such numbers did not resort thither as was usual. Many of the clergymen likewise were dead, and others gone into the country; for it really required a steady courage and a strong faith for a man not only to venture being in town at such a time as this, but likewise to venture to come to church and perform the office of a minister to a congregation, of whom he had reason to believe many of them were actually infected with the plague, and to do this every day, or twice a day, as in some places was done.

It is true the people showed an extraordinary zeal in these religious exercises, and as the church-doors were always open, people would go in single at all times, whether the minister was officiating or no, and locking themselves into separate pews, would be praying to God with great fervency and devotion.

Others assembled at meeting-houses, every one as their different opinions in such things guided, but all were promiscuously the subject of these men's drollery, especially at the beginning of the visitation.

It seems they had been checked for their open insulting religion in this manner by several good people of every persuasion, and that, and the violent raging of the infection, I suppose, was the occasion that they had abated much of their rudeness for some time before, and were only roused by the spirit of ribaldry and atheism at the clamour which was made when the gentleman was first brought in there, and perhaps were agitated by the same devil, when I took upon me to reprove them; though I did it at first with all the calmness, temper, and good manners that I could, which for a while they insulted me the more for thinking it had been in fear of their resentment, though afterwards they found the contrary.

I went home, indeed, grieved and afflicted in my mind at the abominable wickedness of those men, not doubting, however, that they would be made dreadful examples of God's justice; for I looked upon this dismal time to be a particular season of Divine vengeance, and that God would on this occasion single out the proper objects of His displeasure in a more especial and remarkable manner than at another time; and that though I did believe that many good people would, and did, fall in the common calamity, and that it was no certain rule to ' judge of the eternal state of any one by their being distinguished in such a time of general destruction neither one way or other; yet, I say, it could not but seem reasonable to believe that God would not think fit to spare by His mercy such open declared enemies, that should insult His name and Being, defy His vengeance, and mock at His worship and worshippers at such a time; no, not though His mercy had thought fit to bear with and spare them at other times; that this was a day of visitation, a day of God's anger, and those words came into my thought, Jer. v. 9: 'Shall I not visit for these things? saith the Lord: and shall not My soul be avenged of such a nation as this?'

These things, I say, lay upon my mind, and I went home very much grieved and oppressed with the horror of these men's wickedness, and to think that anything could be so vile, so hardened, and notoriously wicked as to insult God, and His servants, and His worship in such a manner, and at such a time as this was, when He had, as it were. His sword drawn in His hand on purpose to take vengeance not on them only, but on the whole nation.

I had, indeed, been in some passion at first with them—though it was really raised, not by any affront they had offered me personally, but by the horror their blaspheming tongues filled me with. However, I was doubtful in my thoughts whether the resentment I retained was not all upon my own private account, for they had given me a great deal of ill language too—I mean personally; but after some pause, and having a weight of grief upon my mind, I retired myself as soon as I came home, for I slept not that night; and giving God most humble thanks for my preservation in the eminent danger I had been in, I set my mind seriously and with the utmost earnestness to pray for those desperate wretches, that God would pardon them, open their eyes, and effectually humble them.

By this I not only did my duty, namely, to pray for those who despitefully used me, but I fully tried my own heart, to my fun satisfaction, that it was not filled with any spirit of resentment as they had offended me in particular; and I humbly recommend the method to all those that would know, or be certain, how to distinguish between their zeal for the honour of God and the effects of their private passions and resentment.

But I must go back here to the particular incidents which occur to my thoughts of the time of the visitation, and particularly to the time of their shutting up houses in the first part of their sickness; for before the sickness was come to its height people had more room to make their observations than they had afterward; but when it was in the extremity there was no such thing as communication with one another, as before.

During the shutting up of houses, as I have said, some violence was offered to the watchmen. As to soldiers, there were none to be found. The few guards which the king then had, which were nothing like the number entertained since, were dispersed, either at Oxford with the Court, or in quarters in the remoter parts of the country, small detachments excepted, who did duty at the Tower and at Whitehall, and these but very few. Neither am I positive that there was any other guard at the Tower than the warders, as they called them, who stand at the gate with gowns and caps, the same as the yeomen of the guard, except the ordinary gunners, who were twenty- four, and the officers appointed to look after the magazine, who were called armourers. As to trained bands, there was no possibility of raising any; neither, if the Lieutenancy, either of London or Middlesex, had ordered the drums to beat for the militia, would any of the companies, I believe, have drawn together, whatever risk they had run.

This made the watchmen be the less regarded, and perhaps occasioned the greater violence to be used against them. I mention it on this score to observe that the setting watchmen thus to keep the people in was, first of all, not effectual, but that the people broke out, whether by force or by stratagem, even almost as often as they pleased; and, second, that those that did thus break out were generally people infected who, in their desperation, running about from one place to another, valued not whom they injured: and which perhaps, as I have said, might give birth to report that it was natural to the infected people to desire to infect others, which report was really false.

And I know it so well, and in so many several cases, that I could give several relations of good, pious, and religious people who, when they have had the distemper, have been so far from being forward to infect others that they have forbid their own family to come near them, in hopes of their being preserved, and have even died without seeing their nearest relations lest they should be instrumental to give them the distemper, and infect or endanger them. If, then, there were cases wherein the infected people were careless of the injury they did to others, this was certainly one of them, if not the chief, namely, when people who had the distemper had broken out from houses which were so shut up, and having been driven to extremities for provision or for entertainment, had endeavoured to conceal their condition, and have been thereby instrumental involuntarily to infect others who have been ignorant and unwary.

This is one of the reasons why I beheved then, and do believe still, that the shutting up houses thus by force, and restraining, or rather imprisoning, people in their own houses, as I said above, was of little or no service in the whole. Nay, I am of opinion it was rather hurtful, having forced those desperate people to wander abroad with the plague upon them, who would otherwise have died quietly in their beds.

I remember one citizen who, having thus broken out of his house in Aldersgate Street or thereabout, went along the road to Islington; he attempted to have gone in at the Angel Inn, and after that the White Horse, two inns known still by the same signs, but was refused; after which he came to the Pied Bull, an inn also still continuing the same sign. He asked them for lodging for one night only, pretending to be going into Lincolnshire, and assuring them of his being very sound and free from the infection, which also at that time had not reached much that way.

They told him they had no lodging that they could spare but one bed up in the garret, and that they could spare that bed for one night, some drovers being expected the next day with cattle; so, if he would accept of that lodging, he might have it, which he did. So a servant was sent up with a candle with him to show him the room. He was very well dressed, and looked like a person not used to lie in a garret; and when he came to the room he fetched a deep sigh, and said to the servant, I have seldom lain in such a lodging as this. 'However, the servant assuring him again that they had no better, 'Well,' says he, 'I must make shift; this is a dreadful time; but it is but for one night.' So he sat down upon the bedside, and bade the maid, I think it was, fetch him up a pint of warm ale. Accordingly the servant went for the ale, but some hurry in the house, which perhaps employed her other ways, put it out of her head, and she went up no more to him.

The next morning, seeing no appearance of the gentleman, somebody in the house asked the servant that had showed him upstairs what was become of him. She started. 'Alas 1' says she, 'I never thought more of him. He bade me carry him some warm ale, but I forgot.' Upon which, not the maid, but some other person was sent up to see after him, who, coming into the room, found him stark dead and almost cold, stretched out across the bed. His clothes were pulled off, his jaw fallen, his eyes open in a most frightful posture, the rug of the bed being grasped hard in one of his hands, so that it was plain he died soon after the maid left him; and 'tis probable, had she gone up with the ale, she had found him dead in a few minutes after he sat down upon the bed. The alarm was great in the house, as anyone may suppose, they having been free from the distemper till that disaster, which, bringing the infection to the house, spread it immediately to other houses round about it. I do not remember how many died in the house itself, but I think the maid-servant who went up first with him fell presently ill by the fright, and several others; for, whereas there died but two in Islington of the plague the week before, there died seventeen the week after, whereof fourteen were of the plague. This was in the week from the 11th of July to the 18th.

There was one shift that some families had, and that not a few, when their houses happened to be infected, and that was this: the families who, in the first breaking-out of the distemper, fled away into the country and had retreats among their friends, generally found some or other of their neighbours or relations to commit the charge of those houses to for the safety of the goods and the like. Some houses were, indeed, entirely locked up, the doors padlocked, the windows and doors having deal boards nailed over them, and only the inspection of them committed to the ordinary watchmen and parish officers; bat these were but few.

It was thought that there were not less than 10,000 houses forsaken of the inhabitants in the city and suburbs, including what was in the out-parishes and in Surrey, or the side of the water they called Southwark. This was besides the numbers of lodgers, and of particular persons who were fled out of other families; so that in all it was computed that about 200,000 people were fled and gone. But of this I shall speak again. But I mention it here on this account, namely, that it was a rule with those who had thus two houses in their keeping or care, that if anybody was taken sick in a family, before the master of the family let the examiners or any other officer know of it, he immediately would send all the rest of his family, whether children or servants, as it fell out to be, to such other house which he had so in charge, and then giving notice of the sick person to the examiner, have a nurse or nurses appointed, and have another person to be shut up in the house with them (which many for money would do), so to take charge of the house in case the person should die.

This was, in many cases, the saving a whole family, who, if they had been shut up with the sick person, would inevitably have perished. But, on the other hand, this was another of the inconveniences of shutting up houses; for the apprehensions and terror of being shut up made many run away with the rest of the family, who, though it was not publicly known, and they were not quite sick, had yet the distemper upon them; and who, by having an uninterrupted liberly to go about, but being obliged still to conceal their circumstances, or perhaps not knowing it themselves, gave the distemper to others, and spread the infection in a dreadful manner, as I shall explain further hereafter.

And here I may be able to make an observation or two of my own, which may be of use hereafter to those into whose hands these may come, if they should ever see the like dreadful visitation.

(1) The infection generally came into the houses of the citizens by the means of their servants, whom they were obliged to send up and down the streets for necessaries; that is to say, for food or physic, to bakehouses, brew-houses, shops, &c.; and who going necessarily through the streets into shops, markets, and the like, it was impossible but that they should, one way or other, meet with distempered people, who conveyed the fatal breath into them, and they brought it home to the families to which they belonged.

(2) It was a great mistake that such a great city as this had but one pest- house; for had there been, instead of one pest-house—viz., beyond Bunhill Fields, where, at most, they could receive, perhaps, two hundred or three hundred people—I say, had there, instead of that one, been several pest- houses, every one able to contain a thousand people, without lying two in a bed, or two beds in a room; and had every master of a family, as soon as any servant especially had been taken sick in his house, been obliged to send them to the next pest-house, if they were willing, as many were, and had the examiners done the like among the poor people when any had been stricken with the infection; I say, had this been done where the people were willing (not otherwise), and the houses not been shut, I am persuaded, and was all the while of that opinion, that not so many, by several thousands, had died; for it was observed, and I could give several instances within the compass of my own knowledge, where a servant had been taken sick, and the family had either time to send him out or retire from the house and leave the sick person, as I have said above, they had all been preserved; whereas when, upon one or more sickening in a family, the house has been shut up, the whole family have perished, and the bearers been obliged to go in to fetch out the dead bodies, not being able to bring them to the door, and at last none left to do it.

(3) This put it out of question to me, that the calamity was spread by infection; that is to say, by some certain steams or fumes, which the physicians call effluvia, by the breath, or by the sweat, or by the stench of the sores of the sick persons, or some other way, perhaps, beyond even the reach of the physicians themselves, which effluvia affected the sound who came within certain distances of the sick, immediately penetrating the vital parts of the said sound persons, putting their blood into an immediate ferment, and agitating their spirits to that degree which it was found they were agitated; and so those newly infected persons communicated it in the same manner to others. And this I shall give some instances of, that cannot but convince those who seriously consider it; and I cannot but with some wonder find some people, now the contagion is over, talk of its being an immediate stroke from Heaven, without the agency of means, having commission to strike this and that particular person, and none other—which I look upon with contempt as the effect of manifest ignorance and enthusiasm; likewise the opinion of others, who talk of infection being carried on by the air only, by carrying with it vast numbers of insects and invisible creatures, who enter into the body with the breath, or even at the pores with the air, and there generate or emit most acute poisons, or poisonous ovae or eggs, which mingle themselves with the blood, and so infect the body: a discourse full of learned simplicity, and manifested to be so by universal experience; but I shall say more to this case in its order.

I must here take further notice that nothing was more fatal to the inhabitants of this city than the supine negligence of the people themselves, who, during the long notice or warning they had of the visitation, made no provision for it by laying in store of provisions, or of other necessaries, by which they might have lived retired and within their own houses, as I have observed others did, and who were in a great measure preserved by that caution; nor were they, after they were a little hardened to it, so shy of conversing with one another, when actually infected, as they were at first: no, though they knew it.

I acknowledge I was one of those thoughtless ones that had made so little provision that my servants were obliged to go out of doors to buy every trifle by penny and halft)enny, just as before it began, even till my experience showing me the folly, I began to be wiser so late that I had scarce time to store myself sufficient for our common subsistence for a month.

I had in family only an ancient woman that managed the house, a maidservant, two apprentices, and myself; and the plague beginning to increase about us, I had many sad thoughts about what course I should take, and how I should act. The many dismal objects which happened everywhere as I went about the streets, had filled my mind with a great deal of horror for fear of the distemper, which was indeed very horrible in itself, and in some more than in others. The swellings, which were generally in the neck or groin, when they grew hard and would not break, grew so painful that it was equal to the most exquisite torture; and some, not able to bear the torment, threw themselves out at windows or shot themselves, or otherwise made themselves away, and I saw several dismal objects of that kind. Others, unable to contain themselves, vented their pain by incessant roarings, and such loud and lamentable cries were to be heard as we walked along the streets that would pierce the very heart to think of, especially when it was to be considered that the same dreadful scourge might be expected every moment to seize upon ourselves.

I cannot say but that now I began to faint in my resolutions; my heart failed me very much, and sorely I repented of my rashness. When I had been out, and met with such terrible things as these I have talked of, I say I repented my rashness in venturing to abide in town. I wished often that I had not taken upon me to stay, but had gone away with my brother and his family.

Terrified by those frightful objects, I would retire home sometimes and resolve to go out no more; and perhaps I would keep those resolutions for three or four days, which time I spent in the most serious thankfulness for my preservation and the preservation of my family, and the constant confession of my sins, giving myself up to God every day, and applying to Him with fasting, humiliation, and meditation. Such intervals as I had I employed in reading books and in writing down my memorandums of what occurred to me every day, and out of which afterwards I took most of this work, as it relates to my observations without doors. What I wrote of my private meditations I reserve for private use, and desire it may not be made public on any account whatever.

I also wrote other meditations upon divine subjects, such as occurred to me at that time and were profitable to myself, but not fit for any other view, and therefore I say no more of that.

I had a very good friend, a physician, whose name was Heath, whom I frequently visited during this dismal time, and to whose advice I was very much obliged for many things which he directed me to take, by way of preventing the infection when I went out, as he found I frequently did, and to hold in my mouth when I was in the streets. He also came very often to see me, and as he was a good Christian as well as a good physician, his agreeable conversation was a very great support to me in the worst of this terrible time.

It was now the beginning of August, and the plague grew very violent and terrible in the place where I lived, and Dr Heath coming to visit me, and finding that I ventured so often out in the streets, earnestly persuaded me to lock myself up and my family, and not to suffer any of us to go out of doors; to keep all our windows fast, shutters and curtains close, and never to open them; but first, to make a very strong smoke in the room where the window or door was to be opened, with rozen and pitch, brimstone or gunpowder and the like; and we did this for some time; but as I had not laid in a store of provision for such a retreat, it was impossible that we could keep within doors entirely. However, I attempted, though it was so very late, to do something towards it; and first, as I had convenience both for brewing and baking, I went and bought two sacks of meal, and for several weeks, having an oven, we baked all our own bread; also I bought malt, and brewed as much beer as all the casks I had would hold, and which seemed enough to serve my house for five or six weeks; also I laid in a quantity of salt butter and Cheshire cheese; but I had no flesh-meat, and the plague raged so violently among the butchers and slaughter-houses on the other side of our street, where they are known to dwell in great numbers, that it was not advisable so much as to go over the street among them.

And here I must observe again, that this necessity of going out of our houses to buy provisions was in a great measure the ruin of the whole city, for the people catched the distemper on these occasions one of another, and even the provisions themselves were often tainted; at least I have great reason to believe so; and therefore I cannot say with satisfaction what I know is repeated with great assurance, that the market-people and such as brought provisions to town were never infected. I am certain the butchers of Whitechappel, where the greatest part of the flesh-meat was killed, were dreadfully visited, and that at least to such a degree that few of their shops were kept open, and those that remained of them killed their meat at Mile End and that way, and brought it to market upon horses.

However, the poor people could not lay up provisions, and there was a necessity that they must go to market to buy, and others to send servants or their children; and as this was a necessity which renewed itself daily, it brought abundance of unsound people to the markets, and a great many that went thither sound brought death home with them.

It is true people used all possible precaution. When any one bought a joint of meat in the market they would not take it off the butcher's hand, but took it off the hooks themselves. On the other hand, the butcher would not touch the money, but have it put into a pot full of vinegar, which he kept for that purpose. The buyer carried always small money to make up any odd sum, that they might take no change. They carried bottles of scents and perfumes in their hands, and all the means that could be used were used, but then the poor could not do even these things, and they went at all hazards.

Innumerable dismal stories we heard every day on this very account. Sometimes a man or woman dropped down dead in the very markets, for many people that had the plague upon them knew nothing of it till the inward gangrene had affected their vitals, and they died in a few moments. This caused that many died frequently in that manner in the streets suddenly, without any warning; others perhaps had time to go to the next bulk or stall, or to any door-porch, and just sit down and die, as I have said before.

These objects were so frequent in the streets that when the plague came to be very raging on one side, there was scarce any passing by the streets but that several dead bodies would be lying here and there upon the ground. On the other hand, it is observable that though at first the people would stop as they went along and call to the neighbours to come out on such an occasion, yet afterward no notice was taken of them; but that if at any time we found a corpse lying, go across the way and not come near it; or, if in a narrow lane or passage, go back again and seek some other way to go on the business we were upon; and in those cases the corpse was always left till the officers had notice to come and take them away, or till night, when the bearers attending the dead-cart would take them up and carry them away. Nor did those undaunted creatures who performed these offices fail to search their pockets, and sometimes strip off their clothes if they were well dressed, as sometimes they were, and carry off what they could get.

But to return to the markets. The butchers took that care that if any person died in the market they had the officers always at band to take them up upon hand-barrows and carry them to the next churchyard; and this was so frequent that such were not entered in the weekly bill, 'Found dead in the streets or fields', as is the case now, but they went into the general articles of the great distemper.

But now the fury of the distemper increased to such a degree that even the markets were but very thinly furnished with provisions or frequented with buyers compared to what they were before; and the Lord Mayor caused the country people who brought provisions to be stopped in the streets leading into the town, and to sit down there with their goods, where they sold what they brought, and went immediately away; and this encouraged the country people greatly-to do so, for they sold their provisions at the very entrances into the town, and even in the fields, as particularly in the fields beyond Whitechappel, in Spittlefields; also in St George's Fields in Southwark, in Bunhill Fields, and in a great field called Wood's Close, near Islington. Thither the Lord Mayor, aldermen, and magistrates sent their officers and servants to buy for their families, themselves keeping within doors as much as possible, and the like did many other people; and after this method was taken the country people came with great cheerfulness, and brought provisions of all sorts, and very seldom got any harm, which, I suppose, added also to that report of their being miraculously preserved.

As for my little family, having thus, as I have said, laid in a store of bread, butter, cheese, and beer, I took my friend and physician's advice, and locked myself up, and my family, and resolved to suffer the hardship of living a few months without flesh-meat, rather than to purchase it at the hazard of our lives.

But though I confined my family, I could not prevail upon my unsatisfied curiosity to stay within entirely myself; and though I generally came frighted and terrified home, yet I could not restrain; only that indeed I did not do it so frequently as at first.

I had some little obligations, indeed, upon me to go to my brother's house, which was in Coleman Street parish and which he had left to my care, and I went at first every day, but afterwards only once or twice a week.

In these walks I had many dismal scenes before my eyes, as particularly of persons falling dead in the streets, terrible shrieks and screechings of women, who, in their agonies, would throw open their chamber windows and cry out in a dismal, surprising manner. It is impossible to describe the variety of postures in which the passions of the poor people would express themselves.

Passing through Tokenhouse Yard, in Lothbury, of a sudden a casement violently opened just over my head, and a woman gave three frightful screeches, and then cried, 'Oh! death, death, death!' in a most inimitable tone, and which struck me with horror and a chillness in my very blood. There was nobody to be seen in the whole street, neither did any other window open, for people had no curiosity now in any case, nor could anybody help one another, so I went on to pass into Bell Alley.

Just in Bell Alley, on the right hand of the passage, there was a more terrible cry than that, though it was not so directed out at the window; but the whole family was in a terrible fright, and I could hear women and children run screaming about the rooms like distracted, when a garret- window opened and somebody from a window on the other side the alley called and asked,

'What is the matter?' upon which, from the first window, it was answered, 'Oh Lord, my old master has hanged himself!' The other asked again, 'Is he quite dead?' and the first answered, 'Ay, ay, quite dead; quite dead and cold!' This person was a merchant and a deputy alderman, and very rich. I care not to mention the name, though I knew his name too, but that would be an hardship to the family, which is now flourishing again.

But this is but one; it is scarce credible what dreadful cases happened in particular families every day. People in the rage of the distemper, or in the torment of their swellings, which was indeed intolerable, running out of their own government, raving and distracted, and oftentimes laying violent hands upon themselves, throwing themselves out at their windows, shooting themselves.,;, &c.; mothers murdering their own children in their lunacy, some dying of mere grief as a passion, some of mere fright and surprise without any infection at all, others frighted into idiotism and foolish distractions, some into despair and lunacy, others into melancholy madness.

The pain of the swelling was in particular very violent, and to some intolerable; the physicians and surgeons may be said to have tortured many poor creatures even to death. The swellings in some grew hard, and they applied violent drawing-plaisters or poultices to break them, and if these did not do they cut and scarified them in a terrible manner. In some those swellings were made hard partly by the force of the distemper and partly by their being too violently drawn, and were so hard that no instrument could cut them, and then they burnt them with caustics, so that many died raving mad with the torment, and some in the very operation. In these distresses, some, for want of help to hold them down in their beds, or to look to them, laid hands upon themselves as above. Some broke out into the streets, perhaps naked, and would run directly down to the river if they were not stopped by the watchman or other officers, and plunge themselves into the water wherever they found it.

It often pierced my very soul to hear the groans and cries of those who were thus tormented, but of the two this was counted the most promising particular in the whole infection, for if these swellings could be brought to a head, and to break and run, or, as the surgeons call it, to digest, the patient generally recovered; whereas those who, like the gentlewoman's daughter, were struck with death at the beginning, and had the tokens come out upon them, often went about indifferent easy till a little before they died, and some till the moment they dropped down, as in apoplexies and epilepsies is often the case. Such would be taken suddenly very sick, and would run to a bench or bulk, or any convenient place that offered itself, or to their own houses if possible, as I mentioned before, and there sit down, grow faint, and die. TTiis kind of dying was much the same as it was with those who die of common mortifications, who die swooning, and, as it were, go away in a dream. Such as died thus had very little notice of their being infected at all till the gangrene was spread through their whole body; nor could physicians themselves know certainly how it was with them till they opened their breasts or other parts of their body and saw the tokens.

We had at this time a great many frightful stories told us of nurses and watchmen who looked after the dying people; that is to say, hired nurses who attended infected people, using them barbarously, starving them, smothering them, or by other wicked means hastening their end, that is to say, murdering of them; and watchmen, being set to guard houses that were shut up when there has been but one person left, and perhaps that one lying sick, that they have broke in and murdered that body, and immediately thrown them out into the dead-cart! And so they have gone scarce cold to the grave.

I cannot say but that some such murders were committed, and I think two were sent to prison for it, but died before they could be tried; and I have heard that three others, at several times, were excused for murders of that kind; but I must say I believe nothing of its being so common a crime as some have since been pleased to say, nor did it seem to be so rational where the people were brought so low as not to be able to help themselves, for such seldom recovered, and there was no temptation to commit a murder, at least none equal to the fact, where they were sure persons would die in so short a time, and could not live.

That there were a great many robberies and wicked practices committed even in this dreadful time I do not deny. The power of avarice was so strong in some that they would run any hazard to steal and to plunder; and particularly in houses where all the families or inhabitants have been dead and carried out, they would break in at all hazards, and without regard to the danger of infection, take even the clothes off the dead bodies and the bed-clothes from others where they lay dead.

This, I suppose, must be the case of a family in Houndsditch, where a man and his daughter, the rest of the family being, as I suppose, carried away before by the dead-cart, were found stark naked, one in one chamber and one in another, lying dead on the floor, and the clothes of the beds, from whence 'tis supposed they were rolled off by thieves, stolen and carried quite away.

It is indeed to be observed that the women were in all this calamity the most rash, fearless, and desperate creatures, and as there were vast numbers that went about as nurses to tend those that were sick, they committed a great many petty thieveries in the houses where they were employed; and some of them were publicly whipped for it, when perhaps they ought rather to have been hanged for examples, for numbers of houses were robbed on these occasions, till at length the parish officers were sent to recommend nurses to the sick, and always took an account whom it was they sent, so as that they might call them to account if the house had been abused where they were placed.

But these robberies extended chiefly to wearing-clothes, linen, and what rings or money they could come at when the person died who was under their care, but not to a general plunder of the houses; and I could give you an account of one of these nurses, who, several years after, being on her deathbed, confessed with the utmost horror the robberies she had committed at the time of her being a nurse, and by which she had enriched herself to a great degree. But as for murders, I do not find that there was ever any proof of the facts in the manner as it has been reported, except as above.

They did tell me, indeed, of a nurse in one place that laid a wet cloth upon the face of a dying patient whom she tended, and so put an end to his life, who was just expiring before; and another that smothered a young woman she was looking to when she was in a fainting fit, and would have come to herself; some that killed them by giving them one thing, some another, and some starved them by giving them nothing at all. But these stories had two marks of suspicion that always attended them, which caused me always to slight them and to look on them as mere stories that people continually frighted one another with. First, that wherever it was that we heard it, they always placed the scene at the farther end of the town, opposite or most remote from where you were to hear it. If you heard it in Whitechappel, it had happened at St Giles's, or at Westminster, or Holbom, or that end of the town. If you heard of it at that end of the town, then it was done in Whitechappel, or the Minories, or about Cripplegate parish. If you heard of it in the city, why, then it happened in Southwark; and if you heard of it in Southwark, then it was done in the city, and the like.

In the next place, of what part soever you heard the story, the particulars were always the same, especially that of laying a wet double clout on a dying man's face, and that of smothering a young gentlewoman; so that it was apparent, at least to my judgement, that there was more of tale than of truth in those things.

However, I cannot say but it had some effect upon the people, and particularly that, as I said before, they grew more cautious whom they took into their houses, and whom they trusted their lives with, and had them always recommended if they could; and where they could not find such, for they were not very plenty, they applied to the parish officers.

But here again the misery of that time lay upon the poor who, being infected, had neither food or physic, neither physician or apothecary to assist them, or nurse to attend them. Many of those died calling for help.

and even for sustenance, out at their windows in a most miserable and deplorable manner; but it must be added that whenever the cases of such persons or families were represented to my Lord Mayor they always were relieved.

It is true, in some houses where the people were not very poor, yet where they had sent perhaps their wives and children away, and if they had any servants they had been dismissed;—I say it is true that to save the expenses, many such as these shut themselves in, and not having help, died alone.

A neighbour and acquaintance of mine, having some money owing to him from a shopkeeper in Whitecross Street or thereabouts, sent his apprentice, a youth about eighteen years of age, to endeavour to get the money. He came to the door, and finding it shut, knocked pretty hard; and, as he thought, heard somebody answer within, but was not sure, so he waited, and after some stay knocked again, and then a third time, when he heard somebody coming downstairs.

At length the man of the house came to the door; he had on his breeches or drawers, and a yellow flannel waistcoat, no stockings, a pair of slipped-shoes, a white cap on his head, and, as the young man said, 'death in his face'.

When he opened the door, says he, 'What do you disturb me thus for?' The boy, though a little surprised, replied, 'I come from such a one, and my master sent me for the money which he says you know of.' 'Very well, child,' returns the living ghost; 'call as you go by at Cripplegate Church, and bid them ring the bell'; and with these words shut the door again, and went up again, and died the same day; nay, perhaps the same hour. This the young man told me himself, and I have reason to believe it. This was while the plague was not come to a height. I think it was in June, towards the latter end of the month; it must be before the dead-carts came about, and while they used the ceremony of ringing the bell for the dead, which was over for certain, in that parish at least, before the month of July, for by the 25th of July there died 550 and upwards in a week, and then they could no more bury in form, rich or poor.

I have mentioned above that notwithstanding this dreadful calamity, yet the numbers of thieves were abroad upon all occasions, where they had found any prey, and that these were generally women. It was one morning about eleven O'clock, I had walked out to my brother's house in Coleman Street parish, as I often did, to see that all was safe.

My brother's house had a little court before it, and a brick wall and a gate in it, and within that several warehouses where his goods of several sorts lay. It happened that in one of these warehouses were several packs of women's high-crowned hats, which came out of the country and were, as I suppose, for exportation: whither, I know not.

I was surprised that when I came near my brother's door, which was in a place they called Swan Alley, I met three or four women with high-crowned hats on their heads; and, as I remembered afterwards, one, if not more, had some hats likewise in their hands; but as I did not see them come out at my brother's door, and not knowing that my brother had any such goods in his warehouse, I did not offer to say anjdhing to them, but went across the way to shun meeting them, as was usual to do at that time, for fear of the plague. But when I came nearer to the gate I met another woman with more hats come out of the gate. 'What business, mistress,' said I, 'have you had there?' There are more people there,' said she; 'I have had no more business there than they.' I was hasty to get to the gate then, and said no more to her, by which means she got away. But just as I came to the gate, I saw two more coming across the yard to come out with hats also on their heads and under their arms, at which I threw the gate to behind me, which having a spring lock fastened itself; and turning to the women, 'Forsooth,' said I, 'what are you doing here?' and seized upon the hats, and took them from them. One of them, who, I confess, did not look like a thief—'Indeed,' says she, 'we are wrong, but we were told they were goods that had no owner. Be pleased to take them again; and look yonder, there are more such customers as we.' She cried and looked pitifully, so I took the hats from her and opened the gate, and bade them be gone, for I pitied the women indeed; but when I looked towards the warehouse, as she directed, there were six or seven more, all women, fitting themselves with hats as unconcerned and quiet as if they had been at a hatter's shop buying for their money.

I was surprised, not at the sight of so many thieves only, but at the circumstances I was in; being now to thrust myself in among so many people, who for some weeks had been so shy of myself that if I met anybody in the street I would cross the way from them.

They were equally surprised, though on another account. They all told me they were neighbours, that they had heard anyone might take them, that they were nobody's goods, and the like. I talked big to them at first, went back to the gate and took out the key, so that they were all my prisoners, threatened to lock them all into the warehouse, and go and fetch my Lord Mayor's officers for them.

They begged heartily, protested they found the gate open, and the warehouse door open; and that it had no doubt been broken open by some who expected to find goods of greater value: which indeed was reasonable to believe, because the lock was broke, and a padlock that hung to the door on the outside also loose, and not abundance of the hats carried away.

At length I considered that this was not a time to be cruel and rigorous; and besides that, it would necessarily oblige me to go much about, to have several people come to me, and I go to several whose circumstances of health I knew nothing of; and that even at this time the plague was so high as that there died 4000 a week; so that in showing my resentment, or even in seeking justice for my brother's goods, I might lose my own life; so I contented myself with taking the names and places where some of them lived, who were really inhabitants in the neighbourhood, and threatening that my brother should call them to an account for it when he returned to his habitation.

Then I talked a little upon another foot with them, and asked them how they could do such things as these in a time of such general calamity, and, as it were, in the face of God's most dreadful judgements, when the plague was at their very doors, and, it may be, in their very houses, and they did not know but that the dead-cart might stop at their doors in a few hours to carry them to their graves.

I could not perceive that my discourse made much impression upon them all that while, till it happened that there came two men of the neighbourhood, hearing of the disturbance, and knowing my brother, for they had been both dependents upon his family, and they came to my assistance. These being, as I said, neighbours, presently knew three of the women and told me who they were and where they lived; and it seems they had given me a true account of themselves before.

This brings these two men to a further remembrance. The name of one was John Hayward, who was at that time undersexton of the parish of St Stephen, Coleman Street. By undersexton was understood at that time gravedigger and bearer of the dead. This man carried, or assisted to carry, all the dead to their graves which were buried in that large parish, and who were carried in form; and after that form of burying was stopped, went with the dead-cart and the bell to fetch the dead bodies from the houses where they lay, and fetched many of them out of the chambers and houses; for the parish was, and is still, remarkable particularly, above all the parishes in London, for a great number of alleys and thoroughfares, very long, into which no carts could come, and where they were obliged to go and fetch the bodies a very long way; which alleys now remain to witness it, such as White's Alley, Cross Key Court, Swan Alley, Bell Alley, White Horse Alley, and many more. Here they went with a kind of hand-barrow and laid the dead bodies on it, and carried them out to the carts; which work he performed and never had the distemper at all, but lived about twenty years after it, and was sexton of the parish to the time of his death. His wife at the same time was a nurse to infected people, and tended many that died in the parish, being for her honesty recommended by the parish officers; yet she never was infected neither.

He never used any preservative against the infection, other than holding garlic and rue in his mouth, and smoking tobacco. This I also had from his own mouth. And his wife's remedy was washing her head in vinegar and sprinkling her head-clothes so with vinegar as to keep them always moist, and if the smell of any of those she waited on was more than ordinary offensive, she snuffed vinegar up her nose and sprinkled vinegar upon her head-clothes, and held a handkerchief wetted with vinegar to her mouth.

It must be confessed that though the plague was chiefly among the poor, yet were the poor the most venturous and fearless of it, and went about their employment with a sort of brutal courage; I must call it so, for it was founded neither on religion nor prudence; scarce did they use any caution, but ran into any business which they could get employment in, though it was the most hazardous. Such was that of tending the sick, watching houses shut up, carrying infected persons to the pest-house, and, which was still worse, carrying the dead away to their graves.

It was under this John Hayward's care, and within his bounds, that the story of the piper, with which people have made themselves so merry, happened, and he assured me that it was true. It is said that it was a blind piper; but, as John told me, the fellow was not bhnd, but an ignorant, weak, poor man, and usually walked his rounds about ten o'clock at night and went piping along from door to door, and the people usually took him in at public-houses where they knew him, and would give him drink and victuals, and sometimes farthings; and he in return would pipe and sing and talk simply, which diverted the people; and thus he lived. It was but a very bad time for this diversion while things were as I have told, yet the poor fellow went about as usual, but was almost starved; and when anybody asked how he did he would answer, the dead cart had not taken him yet, but that they had promised to call for him next week.

It happened one night that this poor fellow, whether somebody had given him too much drink or no—John Hayward said he had not drink in his house, but that they had given him a little more victuals than ordinary at a public-house in Coleman Street—and the poor fellow, having not usually had a bellyful for perhaps not a good while, was laid all along upon the top of a bulk or stall, and fast asleep, at a door in the street near London Wall, towards Cripplegate-, and that upon the same bulk or stall the people of some house, in the alley of which the house was a corner, hearing a bell which they always rang before the cart came, had laid a body really dead of the plague just by him, thinking, too, that this poor fellow had been a dead body, as the other was, and laid there by some of the neighbours.

Accordingly, when John Hayward with his bell and the cart came along, finding two dead bodies lie upon the stall, they took them up with the instrument they used and threw them into the cart, and, all this while the piper slept soundly.

From hence they passed along and took in other dead bodies, till, as honest John Hayward told me, they almost buried him alive in the cart; yet all this while he slept soundly. At length the cart came to the place where the bodies were to be thrown into the ground, which, as I do remember, was at Mount Mill; and as the cart usually stopped some time before they were ready to shoot out the melancholy load they had in it, as soon as the cart stopped the fellow awaked and struggled a little to get his head out from among the dead bodies, when, raising himself up in the cart, he called out, 'Hey! where am I?' This frighted the fellow that attended about the work; but after some pause John Hayward, recovering himself, said, 'Lord, bless us! There's somebody in the cart not quite dead!' So another called to him and said, 'Who are you?' The fellow answered, 'I am the poor piper. Where am I?' 'Where are you?' says Hayward. 'Why, you are in the dead-cart, and we are going to bury you.' 'But I ain't dead though, am I?' says the piper, which made them laugh a little though, as John said, they were heartily frighted at first; so they helped the poor fellow down, and he went about his business.

I know the story goes he set up his pipes in the cart and frighted the bearers and others so that they ran away; but John

Hayward did not tell the story so, nor say anything of his piping at all; but that he was a poor piper, and that he was carried away as above I am fully satisfied of the truth of.

It is to be noted here that the dead-carts in the city were not confined to particular parishes, but one cart went through several parishes, according as the number of dead presented; nor were they tied to carry the dead to their respective parishes, but many of the dead taken up in the city were carried to the burying-ground in the out-parts for want of room.

I have already mentioned the surprise that this judgement was at first among the people. I must be allowed to give some of my observations on the more serious and religious part. Surely never city, at least of this bulk and magnitude, was taken in a condition so perfectly unprepared for such a dreadful visitation, whether I am to speak of the civil preparations or religious. They were, indeed, as if they had had no warning, no expectation, no apprehensions, and consequently the least provision imaginable was made for it in a public way. For example, the Lord Mayor and sheriffs had made no provision as magistrates for the regulations which were to be observed. They had gone into no measures for relief of the poor. The citizens had no public magazines or storehouses for corn or meal for the subsistence of the poor, which if they had provided themselves, as in such cases is done abroad, many miserable families who were now reduced to the utmost distress would have been relieved, and that in a better manner than now could be done.

The stock of the city's money I can say but little to. The Chamber of London was said to be exceedingly rich, and it may be concluded that they were so, by the vast of money issued from thence in the rebuilding the public edifices after the fire of London, and in building new works, such as, for the first part, the Guildhall, Blackwell Hall, part of Leadenhall, half the Exchange, the Session House, the Compter, the prisons of Ludgate, Newgate, &c., several of the wharfs and stairs and landing-places on the river; all which were either burned down or damaged by the great fire of London, the next year after the plague; and of the second sort, the Monument, Fleet Ditch with its bridges, and the Hospital of Bethlem or Bedlam, &c. But possibly the managers of the city's credit at that time made more conscience of breaking in upon the orphan's money to show charity to the distressed citizens than the managers in the following years did to beautify the city and re-edify the buildings; though, in the first case, the losers would have thought their fortunes better bestowed, and the public faith of the city have been less subjected to scandal and reproach.

It must be acknowledged that the absent citizens, who, though they were fled for safety into the country, were yet greatly interested in the welfare of those whom they left behind, forgot not to contribute liberally to the relief of the poor, and large sums were also collected among trading towns in the remotest parts of England; and, as I have heard also, the nobility and the gentry in all parts of England took the deplorable condition of the city into their consideration, and sent up large sums of money in charity to the Lord Mayor and magistrates for the relief of the poor. The king also, as I was told, ordered a thousand pounds a week to be distributed in four parts: one quarter to the city and liberty of Westminster; one quarter or part among the inhabitants of the Southwark side of the water; one quarter to the liberty and parts within of the city, exclusive of the city within the walls; and one-fourth part to the suburbs in the county of Middlesex, and the east and north parts of the city. But this latter I only speak of as a report.

Certain it is, the greatest part of the poor or families who formerly lived by their labour, or by retail trade, lived now on charity; and had there not been prodigious sums of money given by charitable, well-minded Christians for the support of such, the city could never have subsisted. There were, no question, accounts kept of their charity, and of the just distribution of it by the magistrates. But as such multitudes of those very officers died through whose hands it was distributed, and also that, as I have been told, most of the accounts of those things were lost in the great fire which happened in the very next year, and which burnt even the chamberlain's office and many of their papers, so I could never come at the particular account, which I used great endeavours to have seen.

It may, however, be a direction in case of the approach of a like visitation, which God keep the city from;—I say, it may be of use to observe that by the care of the Lord Mayor and aldermen at that time in distributing weekly great sums of money for relief of the poor, a multitude of people who would otherwise have perished, were relieved, and their fives preserved. And here let me enter into a brief state of the case of the poor at that time, and what way apprehended from them, from whence may be judged hereafter what may be expected if the like distress should come upon the city.

At the beginning of the plague, when there was now no more hope but that the whole city would be visited; when, as I have said, all that had friends or estates in the country retired with their families; and when, indeed, one would have thought the very city itself was running out of the gates, and that there would be nobody left behind; you may be sure from that hour all trade, except such as related to immediate subsistence, was, as it were, at a full stop.

This is so lively a case, and contains in it so much of the real condition of the people, that I think I cannot be too particular in it, and therefore I descend to the several arrangements or classes of people who fell into immediate distress upon this occasion. For example:

1. All master-workmen in manufactures, especially such as belonged to ornament and the less necessary parts of the people's dress, clothes, and furniture for houses, such as riband-weavers and other weavers, gold and silver lace makers, and gold and silver wire drawers, sempstresses, milliners, shoemakers, hatmakers, and glovemakers; also upholsterers, joiners, cabinet-makers, looking-glass makers, and innumerable trades which depend upon such as these;—I say, the master-workmen in such stopped their work, dismissed their journeymen and workmen, and all their dependents.

2. As merchandising was at a full stop, for very few ships ventured to come up the river and none at all went out, so all the extraordinary officers of the customs, likewise the watermen, cartmen, porters, and all the poor whose labour depended upon the merchants, were at once dismissed and put out of business.

3. All the tradesmen usually employed in building or repairing of houses were at a full stop, for the people were far from wanting to build houses when so many thousand houses were at once stripped of their inhabitants; so that this one article turned all the ordinary workmen of that kind out of business, such as bricklayers, masons, carpenters, joiners, plasterers, painters, glaziers, smiths, plumbers, and all the labourers depending on such.

4. As navigation was at a stop, our ships neither coming in or going out as before, so the seamen were all out of employment, and many of them in the last and lowest degree of distress; and with the seamen were all the several tradesmen and workmen

belonging to and depending upon the building and fitting out of ships, such as ship-carpenters, caulkers, ropemakers, dry coopers, sailmakers, anchorsmiths, and other smiths; blockmakers, carvers, gunsmiths, ship-chandlers, ship-carvers, and the like. The masters of those perhaps might live upon their substance, but the traders were universally at a stop, and consequently all their workmen discharged. Add to these that the river was in a manner without boats, and all or most part of the watermen, lightermen, boat-builders, and lighter-builders in like manner idle and laid by.

5. All families retrenched their living as much as possible, as well those that fled as those that stayed; so that an innumerable multitude of footmen, serving-men, shopkeepers, journeymen, merchants' bookkeepers, and such sort of people, and especially poor maid-servants, were turned off, and left friendless and helpless, without employment and without habitation, and this was really a dismal article.

I might be more particular as to this part, but it may suffice to mention in general, all trades being stopped, employment ceased: the labour, and by that the bread, of the poor were cut off; and at first indeed the cries of the poor were most lamentable to hear, though by the distribution of charity their misery that way was greatly abated. Many indeed fled into the counties, but thousands of them having stayed in London till nothing but desperation sent them away, death overtook them on the road, and they served for no better than the messengers of death; indeed, others carrying the infection along with them, spread it very unhappily into the remotest parts of the kingdom.

Many of these were the miserable objects of despair which I have mentioned before, and were removed by the destruction which followed. These might be said to perish not by the infection itself but by the consequence of it; indeed, namely, by hunger and distress and the want of all things: being without lodging, without money, without friends, without means to get their bread, or without anyone to give it them; for many of them were without what we call legal settlements, and so could not claim of the parishes, and all the support they had was by application to the magistrates for relief, which relief was (to give the magistrates their due) carefully and cheerfully administered as they found it necessary, and those that stayed behind never felt the want and distress of that kind which they felt who went away in the manner above noted.

Let any one who is acquainted with what multitudes of people get their daily bread in this city by their labour, whether artificers or mere workmen—I say, let any man consider what must be the miserable condition of this town if, on a sudden, they should be all turned out of employment, that labour should cease, and wages for work be no more.

This was the case with us at that time; and had not the sums of money contributed in charity by well-disposed people of every kind, as well abroad as at home, been prodigiously great, it had not been in the power of the Lord Mayor and sheriffs to have kept the public peace. Nor were they without apprehensions, as it was, that desperation should push the people upon tumults, and cause them to rifle the houses of rich men and plunder the markets of provisions; in which case the country people, who brought provisions very freely and boldly to town, would have been terrified from coming any more, and the town would have sunk under an unavoidable famine.

But the prudence of my Lord Mayor and the Court of Aldermen within the city, and of the justices of peace in the out-parts, was such, and they were supported with money from all parts so well, that the poor people were kept quiet, and their wants everywhere relieved, as far as was possible to be done.

Two things besides this contributed to prevent the mob doing any mischief. One was, that really the rich themselves had not laid up stores of provisions in their houses as indeed they ought to have done, and which if they had been wise enough to have done, and locked themselves entirely up, as some few did, they had perhaps escaped the disease better. But as it appeared they had not, so the mob had no notion of finding stores of provisions there if they had broken in. as it is plain they were sometimes very near doing, and which: if they bad, they had finished the ruin of the whole city, for there were no regular troops to have withstood them, nor could the trained bands have been brought together to defend the city, no men being to be found to bear arms.

But the vigilance of the Lord Mayor and such magistrates as could be had (for some, even of the aldermen, were dead, and some absent) prevented this; and they did it by the most kind and gentle methods they could think of, as particularly by relieving the most desperate with money, and putting others into business, and particularly that employment of watching houses that were infected and shut up. And as the number of these were very great (for it was said there was at one time ten thousand houses shut up, and every house had two watchmen to guard it, viz., one by night and the other by day), this gave opportunity to employ a very great number of poor men at a time.

The women and servants that were turned off from their places were likewise employed as nurses to tend the sick in all places, and this took off a very great number of them.

And, which though a melancholy article in itself, yet was a deliverance in its kind: namely, the plague, which raged in a dreadful manner from the middle of August to the middle of

October, carried off in that time thirty or forty thousand of these very people which, had they been left, would certainly have been an insufferable burden by their poverty; that is to say, the whole city could not have supported the expense of them, or have provided food for them; and they would in time have been even driven to the necessity of plundering either the city itself or the country adjacent, to have subsisted themselves, which would first or last have put the whole nation, as well as the city, into the utmost terror and confusion.

It was observable, then, that this calamity of the people made them very humble; for now for about nine weeks together there died near a thousand a day, one day with another, even by the account of the weekly bills, which yet, I have reason to be assured, never gave a full account, by many thousands; the confusion being such, and the carts working in the dark when they carried the dead, that in some places no account at all was kept, but they worked on, the clerks and sextons not attending for weeks together, and not knowing what number they carried. This account is verified by the following bills of mortality:—

Of all of the
Diseases. Plague
From August 8 to August 15 5319 3880
" " 15 " 22 5568 4237
" " 22 " 29 7496 6102
" " 29 to September 5 8252 6988
"September 5 " 12 7690 6544
" " 12 " 19 8297 7165
" " 19 " 26 6460 5533
" " 26 to October 3 5720 4979
"October 3 " 10 5068 4327

———————

59,870 49,705

So that the gross of the people were carried off in these two months; for, as the whole number which was brought in to die of the plague was but 68,590, here is 50,000 of them, within a trifle, in two months; I say 50,000, because, as there wants 295 in the number above, so there wants two days of two months in the account of time.

Now when I say that the parish officers did not give in a full account, or were not to be depended upon for their account, let any one but consider how men could be exact in such a time of dreadful distress, and when many of them were taken sick themselves and perhaps died in the very time when their accounts were to be given in; I mean the parish clerks, besides inferior officers; for though these poor men ventured at all hazards, yet they were far from being exempt from the common calamity, especially if it be true that the parish of Stepney had, within the year, 116 sextons, gravediggers. and their assistants; that is to say, bearers, bellmen, and drivers of carts for carrying off the dead bodies.

Indeed the work was not of a nature to allow them leisure to take an exact tale of the dead bodies, which were all huddled together in the dark into a pit; which pit or trench no man could come nigh but at the utmost peril. I observed often that in the parishes of Aldgate and Cripplegate, Whitechappel and Stepney, there were five, six, seven, and eight hundred in a week in the bills; whereas if we may believe the opinion of those that lived in the city all the time as well as I, there died sometimes 2000 a week in those parishes; and I saw it under the hand of one that made as strict an examination into that part as he could, that there really died an hundred thousand people of the plague in that one year whereas in the bills, the articles of the plague, it was but 68,590.

If I may be allowed to give my opinion, by what I saw with my eyes and heard from other people that were eye-witnesses, I do verily believe the same, viz., that there died at least 100,000 of the plague only, besides other distempers and besides those which died in the fields and highways and secret Places out of the compass of the communication, as it was called, and who were not put down in the bills though they really belonged to the body of the inhabitants. It was known to us all that abundance of poor despairing creatures who had the distemper upon them, and were grown stupid or melancholy by their misery, as many were, wandered away into the fields and Woods, and into secret uncouth places almost anywhere, to creep into a bush or hedge and die.

The inhabitants of the villages adjacent would, in pity, carry them food and set it at a distance, that they might fetch it, if they were able; and sometimes they were not able, and the next time they went they should find the poor wretches lie dead and the food untouched. The number of these miserable objects were many, and I know so many that perished thus, and so exactly where, that I believe I could go to the very place and dig their bones up still; for the country people would go and dig a hole at a distance from them, and then with long poles, and hooks at the end of them, drag the bodies into these pits, and then throw the earth in from as far as they could cast it, to cover them, taking notice how the wind blew, and so coming on that side which the seamen call to windward, that the scent of the bodies might blow from them; and thus great numbers went out of the world who were never known, or any account of them taken, as well within the bills of mortality as without.

This, indeed, I had in the main only firom the relation of others, for I seldom walked into the fields, except towards Bethnal Green and Hactoey, or as hereafter. But when I did walk, I always saw a great many poor wanderers at a distance; but I could know httle of their cases, for whether it were in the street or in the fields, if we had seen anybody coming, it was a general method to walk away; yet I believe the account is exactly true.

As this puts me upon mentioning my walking the streets and fields, I cannot omit taking notice what a desolate place the city was at that time. The great street I lived in (which is known to be one of the broadest of all the streets of London, I mean of the suburbs as well as the liberties) all the side where the butchers lived, especially without the bars, was more like a green field than a paved street, and the people generally went in the middle with the horses and carts. It is true that the farthest end towards Whitechappel Church was not all paved, but even the part that was paved was full of grass also; but this need not seem strange, since the great streets within the city, such as Leadenhall Street, Bishopsgate Street, Cornhill, and even the Exchange itself, had grass growing in them in several places; neither cart or coach were seen in the streets from morning to evening, except some country carts to bring roots and beans, or peas, hay, and straw, to the market, and those but very few compared to what was usual. As for coaches, they were scarce used but to carry sick people to the pest-house, and to other hospitals, and some few to carry physicians to such places as they thought fit to venture to visit; for really coaches were dangerous things, and people did not care to venture into them, because they did not know who might have been carried in them last, and sick, infected people were, as I have said, ordinarily carried in them to the pest-houses, and sometimes people expired in them as they went along.

It is true, when the infection came to such a height as I have now mentioned, there were very few physicians which cared to stir abroad to sick houses, and very many of the most eminent of the faculty were dead, as well as the surgeons also; for now it was indeed a dismal time, and for about a month together, not taking any notice of the bills of mortality, I believe there did not die less than 1500 or 1700 a day, one day with another.

One of the worst days we had in the whole time, as I thought, was in the beginning of September, when, indeed, good people began to think that God was resolved to make a full end of the people in this miserable city. This was at that time when the plague was fully come into the eastern parishes. The parish of Aldgate, if I may give my opinion, buried above a thousand a week for two weeks, though the bills did not say so many;—but it surrounded me at so dismal a rate that there was not a house in twenty uninfected in the Minories, in Houndsditch, and in those parts of Aldgate parish about the Butcher Row and the alleys over against me. I say, in those places death reigned in every comer. Whitechappel parish was in the same condition, and though much less than the parish I lived in, yet buried near 600 a week by the bills, and in my opinion near twice as many. Whole famines, and indeed whole streets of families, were swept away together; insomuch that it was frequent for neighbours to call to the belhnan to go to such-and-such houses and fetch out the people, for that they were all dead.

And, indeed, the work of removing the dead bodies by carts was now grown so very odious and dangerous that it was complained of that the bearers did not take care to dear such houses where all the inhabitants were dead, but that sometimes the bodies lay several days unburied, till the neighbouring families were offended with the stench, and consequently infected; and this neglect of the officers was such that the churchwanlens and constables were summoned to look after it, and even the justices of the Hamlets were obliged to venture their lives among them to quicken and encourage them, for innumerable of the bearers died of the distemper, infected by the bodies they were obliged to come so near. And had it not been that the number of poor people who wanted employment and wanted bread (as I have said before) was so great that necessity drove them to undertake anything and venture anything, they would never have found people to be employed. And then the bodies of the dead would have lain above ground, and have perished and rotted in a dreadful manner.

But the magistrates cannot be enough commended in this, that they kept such good order for the burying of the dead, that as fast as any of these they employed to carry off and bury the dead fell sick or died, as was many times the case, they immediately supplied the places with others, which, by reason of the great number of poor that was left out of business, as above, was not hard to do. This occasioned, that notwithstanding the infinite number of people which died and were sick, almost all together, yet they were always cleared away and carried off every night, so that it was never to be said of London that the living were not able to bury the dead.

As the desolation was greater during those terrible times, so the amazement of the people increased, and a thousand unaccountable things they would do in the violence of their fright, as others did the same in the agonies of their distemper, and this part was very affecting. Some went roaring and crying and wringing their hands along the street; some would go praying and lifting up their hands to heaven, calling upon God for mercy. I cannot say, indeed, whether this was not in their distraction, but, be it so, it was still an indication of a more serious mind, when they had the use of their senses, and was much better, even as it was, than the frightful yellings and cryings that every day, and especially in the evenings, were heard in some streets. I suppose the world has heard of the famous Solomon Eagle, an enthusiast. He, though not infected at all but in his head, went about denouncing of judgement upon the city in a frightful manner, sometimes quite naked, and with a pan of burning charcoal on his head. What he said, or pretended, indeed I could not learn.

I will not say whether that clergyman was distracted or not, or whether he did it in pure zeal for the poor people, who went every evening through the streets of Whitechappel, and, with his hands lifted up, repeated that part of the Liturgy of the Church continually, 'Spare us, good Lord; spare Thy people, whom Thou has redeemed with Thy most precious blood.' I say, I cannot speak positively of these things, because these were only the dismal objects which represented themselves to me as I looked through my chamber windows (for I seldom opened the casements), while I confined myself within doors during that most violent raging of the pestilence; when, indeed, as I have said, many began to think, and even to say, that there would none escape; and indeed I began to think so too, and therefore kept within doors for about a fortnight and never stirred out. But I could not hold it. Besides, there were some people who, notwithstanding the danger, did not omit publicly to attend the worship of God, even in the most dangerous times; and though it is true that a great many clergymen did shut up their churches, and fled, as other people did, for the safety of their lives, yet all did not do so. Some ventured to officiate and to keep up the assemblies of the people by constant prayers, and sometimes sermons or brief exhortations to repentance and reformation, and this as long as any would come to hear them. And Dissenters did the like also, and even in the very churches where the parish ministers were either dead or fled; nor was there any room for making difference at such a time as this was.

It was indeed a lamentable thing to hear the miserable lamentations of poor dying creatures calling out for ministers to comfort them and pray with them, to counsel them and to direct them, calling out to God for pardon and mercy, and confessing aloud their past sins. It would make the stoutest heart bleed to hear how many warnings were then given by dying penitents to others not to put off and delay their repentance to the day of distress; that such a time of calamity as this was no time for repentance, was no time to call upon God. I wish I could repeat the very sound of those groans and of those exclamations that I heard from some poor dying creatures when in the height of their agonies and distress, and that I could make him that reads this hear, as I imagine I now hear them, for the sound seems still to ring in my ears.

If I could but tell this part in such moving accents as should alarm the very soul of the reader, I should rejoice that I recorded those things, however short and imperfect.

It pleased God that I was still spared, and very hearty and sound in health, but very impatient of being pent up within doors without air, as I had been for fourteen days or thereabouts; and I could not restrain myself, but I would go to carry a letter for my brother to the post-house. Then it was indeed that I observed a profound silence in the streets. When I came to the post-house, as I went to put in my letter I saw a man stand in one corner of the yard and talking to another at a window, and a third had opened a door belonging to the office. In the middle of the yard lay a small leather purse with two keys hanging at it, with money in it, but nobody would meddle with it. I asked how long it had lain there; the man at the window said it had lain almost an hour, but that they had not meddled with it, because they did not know but the person who dropped it might come back to look for it. I had no such need of money, nor was the sum so big that I had any inclination to meddle with it, or to get the money at the hazard it might be attended with; so I seemed to go away, when the man who had opened the door said he would take it up, but so that if the right owner came for it he should be sure to have it. So he went in and fetched a pail of water and set it down hard by the purse, then went again and fetch some gunpowder, and cast a good deal of powder upon the purse, and then made a train from that which he had thrown loose upon the purse. The train reached about two yards. After this he goes in a third time and fetches out a pair of tongs red hot, and which he had prepared, I suppose, on purpose; and first setting fire to the train of powder, that singed the purse and also smoked the air sufficiently. But he was not content with that, but he then takes up the purse with the tongs, holding it so long till the tongs burnt through the purse, and then he shook the money out into the pail of water, so he carried it in. The money, as I remember, was about thirteen shilling and some smooth groats and brass farthings.

There might perhaps have been several poor people, as I have observed above, that would have been hardy enough to have ventured for the sake of the money; but you may easily see by what I have observed that the few people who were spared were very careful of themselves at that time when the distress was so exceeding great.

Much about the same time I walked out into the fields towards Bow; for I had a great mind to see how things were managed in the river and among the ships; and as I had some concern in shipping, I had a notion that it had been one of the best ways of securing one's self from the infection to have retired into a ship; and musing how to satisfy my curiosity in that point, I turned away over the fields from Bow to Bromley, and down to Blackwall to the stairs which are there for landing or taking water.

Here I saw a poor man walking on the bank, or sea-wall, as they call it, by himself. I walked a while also about, seeing the houses all shut up. At last I fell into some talk, at a distance, with this poor man; first I asked him how people did thereabouts. 'Alas, sir!' says he, 'almost desolate; all dead or sick. Here are very few families in this part, or in that village' (pointing at Poplar), 'where half of them are not dead already, and the rest sick.' Then he pointing to one house. There they are all dead', said he, 'and the house stands open; nobody dares go into it. A poor thief, says he, 'ventured in to steal something, but he paid dear for his theft, for he was carried to the churchyard too last night.' Then he pointed to several other houses. 'There', says he. 'they are all dead, the man and his wife, and five children. There', says he, 'they are shut up; you see a watchman at the door'; and so of other houses. 'Why,' says I, 'what do you here all alone? ' 'Why,' says he, 'I am a poor, desolate man; it has pleased God I am not yet visited, though my family is, and one of my children dead.' 'How do you mean, then,' said I, 'that you are not visited?' 'Why,' says he, 'that's my house' (pointing to a very little, low-boarded house), 'and there my poor wife and two children live,' said he, 'if they may be said to live, for my wife and one of the children are visited, but I do not come at them.' And with that word I saw the tears run very plentifully down his face; and so they did down mine too, I assure you.

'But,' said I, 'why do you not come at them? How can you abandon your own flesh and blood?' 'Oh, sir,' says he, 'the Lord forbid! I do not abandon them; I work for them as much as I am able; and, blessed be the Lord, I keep them from want'; and with that I observed he lifted up his eyes to heaven, with a countenance that presently told me I had happened on a man that was no hypocrite, but a serious, religious, good man, and his ejaculation was an expression of thankfulness that, in such a condition as he was in, he should be able to say his family did not want. 'Well,' says I, 'honest man, that is a great mercy as things go now with the poor. But how do you live, then, and how are you kept from the dreadful calamity that is now upon us all?' 'Why, sir,' says he, 'I am a waterman, and there's my boat,' says he, 'and the boat serves me for a house. I work in it in the day, and I sleep in it in the night; and what I get I lay down upon that stone,' says he, showing me a broad stone on the other side of the street, a good way from his house; 'and then,' says he, 'I halloo, and call to them till I make them hear; and they come and fetch it.'

'Well, friend,' says I, 'but how can you get any money as a waterman? Does an body go by water these times?' 'Tes, sir,' says he, 'in the way I am employed there does. Do you see there,' says he, 'five ships lie at anchor' (pointing down the river a good way below the town), 'and do you see', says he, 'eight or ten ships lie at the chain there, and at anchor yonder?' pointing above the town). 'All those ships have families on board, of their merchants and owners, and such-like, who have locked themselves up and live on board, close shut in, for fear of the infection; and I tend on them to fetch things for them, carry letters, and do what is absolutely necessary, that they may not be obliged to come on shore; and every night I fasten my boat on board one of the ship's boats, and there I sleep by myself, and, blessed be God, I am preserved hitherto.'

'Well,' said I, 'friend, but will they let you come on board after you have been on shore here, when this is such a terrible place, and so infected as it is?'

'Why, as to that,' said he, 'I very seldom go up the ship-side, but deliver what I bring to their boat, or lie by the side, and they hoist it on board. If I did, I think they are in no danger from me, for I never go into any house on shore, or touch anybody, no, not of my own family; but I fetch provisions for them.'

'Nay,' says I, 'but that may be worse, for you must have those provisions of somebody or other; and since all this part of the town is so infected, it is dangerous so much as to speak with anybody, for the village', said I, 'is, as it were, the beginning of London, though it be at some distance from it.'

'That is true,' added he; 'but you do not understand me right; I do not buy provisions for them here. I row up to Greenwich and buy fresh meat there, and sometimes I row down the river to Woolwich and buy there; then I go to single farm-houses on the Kentish side, where I am known, and buy fowls and eggs and butter, and bring to the ships, as they direct me, sometimes one, sometimes the other. I seldom come on shore here, and I came now only to call on my wife and hear how my family do, and give them a little money, which I received last night.'

'Poor man!' said I; 'and how much hast thou gotten for them?'

'I have gotten four shillings,' said he, 'which is a great sum, as things go now with poor men; but they have given me a bag of bread too, and a salt fish and some flesh; so all helps out.' 'Well,' said I, 'and have you given it them yet?'

'No,' said he; 'but I have called, and my wife has answered that she cannot come out yet, but in half-an-hour she hopes to come, and I am waiting for her. Poor woman!' says he, 'she is brought sadly down. She has a swelling, and it is broke, and I hope she will recover; but I fear the child will die, but it is the Lord—'

Here he stopped, and wept very much.

'Well, honest friend,' said I, 'thou hast a sure Comforter, if thou hast brought thyself to be resigned to the will of God; He is dealing with us all in judgement.'

'Oh, sir!' says he, 'it is infinite mercy if any of us are spared, and who am I to repine!'

'Sayest thou so?' said I, 'and how much less is my faith than thine?' And here my heart smote me, suggesting how much better this poor man's foundation was on which he stayed in the danger than mine; that he had nowhere to fly; that he had a family to bind him to attendance, which I had not; and mine was mere presumption, his a true dependence and a courage resting on God; and yet that he used all possible caution for his safety.

I turned a little way from the man while these thoughts engaged me, for, indeed, I could no more refrain from tears than he.

At length, after some further talk, the poor woman opened the door and called, 'Robert, Robert'. He answered, and bid her stay a few moments and he would come; so he ran down the common stairs to his boat and fetched up a sack, in which was the provisions he had brought from the ships; and when he returned he hallooed again. Then he went to the great stone which he showed me and emptied the sack, and laid all out, everything by themselves, and then retired; and his wife came with a little boy to fetch them away, and called and said such a captain had sent such a thing, and such a captain such a thing, and at the end adds, 'God has sent it all; give thanks to Him.' When the poor woman had taken up all, she was so weak she could not carry it at once in, though the weight was not much neither; so she left the biscuit, which was in a little bag, and left a little boy to watch it till she came again.

'Well, but', says I to him, 'did you leave her the four shillings too, which you said was your week's pay?'

'Yes, yes,' says he; 'you shall hear her own it.' So he calls again, 'Rachel, Rachel,' which it seems was her name, 'did you take up the money?' 'Yes,' said she. 'How much was it?' said he. 'Four shillings and a groat,' said she. 'Well, well,' says he, 'the Lord keep you all'; and so he turned to go away.

As I could not refrain contributing tears to this man's story, so neither could I refrain my charity for his assistance. So I called him, 'Hark thee, friend,' said I, 'come hither, for I believe thou art in health, that I may venture thee'; so I pulled out my hand, which was in my pocket before, 'Here,' says I, 'go and call thy Rachel once more, and give her a little more comfort from me. God will never forsake a family that trust in Him as thou dost.' So I gave him four other shillings, and bid him go lay them on the stone and call his wife.

I have not words to express the poor man's thankfulness, neither could he express it himself but by tears running down his face. He called his wife, and told her God had moved the heart of a stranger, upon hearing their condition, to give them all that money, and a great deal more such as that he said to her. The woman, too, made signs of the like thankfulness, as well to Heaven as to me, and joyfully picked it up; and I parted with no money all that year that I thought better bestowed.

I then asked the poor man if the distemper had not reached to Greenwich. He said it had not till about a fortnight before; but that then he feared it had, but that it was only at that end of the town which lay south towards Deptford Bridge; that he went only to a butcher's shop and a grocer's, where he generally bought such things as they sent him for, but was very careful.

I asked him then how it came to pass that those people who had so shut themselves up in the ships had not laid in sufficient stores of all things necessary. He said some of them had—but, on the other hand, some did not come on board till they were frighted into it and till it was too dangerous for them to go to the proper people to lay in quantities of things, and that he waited on two ships, which he showed me, that had laid in little or nothing but biscuit bread and ship beer, and that he had bought everything else almost for them. I asked him if there was any more ships that had separated themselves as those had done. He told me yes, all the way up from the point, right against Greenwich, to within the shore of Limehouse and Redriff, all the ships that could have room rid two and two in the middle of the stream, and that some of them had several families on board. I asked him if the distemper had not reached them. He said he believed it had not, except two or three ships whose people had not been so watchful to keep the seamen from going on shore as others had been, and he said it was a very fine sight to see how the ships lay up the Pool.

When he said he was going over to Greenwich as soon as the tide began to come in, I asked if he would let me go with him and bring me back, for that I had a great mind to see how the ships were ranged, as he had told me. He told me, if I would assure him on the word of a Christian and of an honest man that I had not the distemper, he would. I assured him that I had not; that it had pleased God to preserve me; that I lived in Whitechappel, but was too impatient of being so long within doors, and that I had ventured out so far for the refreshment of a little air, but that none in my house had so much as been touched with it.

Well, sir,' says he, 'as your charity has been moved to pity me and my poor family, sure you cannot have so little pity left as to put yourself into my boat if you were not sound in health which would be nothing less than killing me and ruining my whole family.' The poor man troubled me so much when he spoke of his family with such a sensible concern and in such an affectionate manner, that I could not satisfy myself at first to go at all. I told him I would lay aside my curiosity rather than make him uneasy, though I was sure, and very thankful for it, that I had no more distemper upon me than the freshest man in the world. Well, he would not have me put it off neither, but to let me see how confident he was that I was just to him, now importuned me to go; so when the tide came up to his boat I went in, and he carried me to Greenwich. While he bought the things which he had in his charge to buy, I walked up to the top of the hill under which the town stands, and on the east side of the town, to get a prospect of the river. But it was a surprising sight to see the number of ships which lay in rows, two and two, and some places two or three such lines in the breadth of the river, and this not only up quite to the town, between the houses which we call Ratcliff and Redriff, which they name the Pool, but even down the whole river as far as the head of Long Reach, which is as far as the hills give us leave to see it.

I cannot guess at the number of ships, but I think there must be several hundreds of sail; and I could not but applaud the contrivance: for ten thousand people and more who attended ship affairs were certainly sheltered here from the violence of the contagion, and lived very safe and very easy.

I returned to my own dwelling very well satisfied with my day's journey, and particularly with the poor man; also I rejoiced to see that such little sanctuaries were provided for so many families in a time of such desolation, I observed also that, as the violence of the plague had increased, so the ships which had families on board removed and went farther off, till, as I was told, some went quite away to sea, and put into such harbours and safe roads on the north coast as they could best come at.

But it was also true that all the people who thus left the land and lived on board the ships were not entirely safe from the infection, for many died and were thrown overboard into the river, some in coffins, and some, as I heard, without coffins, whose bodies were seen sometimes to drive up and down with the tide in the river.

But I believe I may venture to say that in those ships which were thus infected it either happened where the people had recourse to them too late, and did not fly to the ship till they had stayed too long on shore and had the distemper upon them (though perhaps they might not perceive it) and so the distemper did not come to them on board the ships, but they really carried it with them; or it was in these ships where the poor waterman said they had not had time to furnish themselves with provisions, but were obliged to send often on shore to buy what they had occasion for, or suffered boats to come to them from the shore. And so the distemper was brought insensibly among them.

And here I cannot but take notice that the strange temper of the people of London at that time contributed extremely to their own destruction. The plague began, as I have observed, at the other end of the town, namely, in Long Acre, Drury Lane, &c., and came on towards the city very gradually and slowly. It was felt at first in December, then again in February, then again in April, and always but a very little at a time; then it stopped till May, and even the last week in May there was but seventeen, and all at that end of the town; and all this while, even so long as till there died above 3000 a week, yet had the people in Redriff, and in Wapping and Ratcliff, on both sides of the river, and almost all Southwark side, a mighty fancy that they should not be visited, or at least that it would not be so violent among them. Some people fancied the smell of the pitch and tar, and such other things as oil and rosin and brimstone, which is so much used by all trades relating to shipping, would preserve them. Others argued it, because it was in its extreamest violence in Westminster and the parish of St Giles and St Andrew, &c., and began to abate again before it came among them—which was true indeed, in part. For example:—

>From the 8th to the 15th August:—
St Giles-in-the-Fields 242
Cripplegate 886
stepney 197
St Margaret, Bermondsey 24
Rotherhith 3
Total this week 4030
From the 15th to the 22nd August:—
St Giles-in-the-Fields 175
Cripplegate 847
Stepney 273
St Margaret, Bermondsey 36
Rotherhith 2
Total this week 5319

N.B.—That it was observed the numbers mentioned in Stepney parish at that time were generally all on that side where Stepney parish joined to Shoreditch, which we now call Spittlefields, where the parish of Stepney comes up to the very wall of Shoreditch Churchyard, and the plague at this time was abated at St Giles-in-the-Fields, and raged most violently in Cripplegate, Bishopsgate, and Shoreditch parishes; but there was not ten people a week that died of it in all that part of Stepney parish which takes in Limehouse, Ratcliff Highway, and which are now the parishes of Shadwell and Wapping, even to St Katherine's by the Tower, till after the whole month of August was expired. But they paid for it afterwards, as I shall observe by-and-by.

This, I say, made the people of Redriff and Wapping, Ratcliff and Limehouse, so secure, and flatter themselves so much with the plague's going off without reaching them, that they took no care either to fly into the country or shut themselves up. Nay, so far were they from stirring that they rather received their friends and relations from the city into their houses, and several from other places really took sanctuary in that part of the town as a Place of safety, and as a place which they thought God would pass over, and not visit as the rest was visited.

And this was the reason that when it came upon them they were more surprised, more unprovided, and more at a loss what to do than they were in other places; for when it came among them really and with violence, as it did indeed in September and October, there was then no stirring out into the country, nobody would suffer a stranger to come near them, no, nor near the towns where they dwelt; and, as I have been told, several that wandered into the country on Surrey side were found starved to death in the woods and commons, that country being more open and more woody than any other part so near London, especially about Norwood and the parishes of Camberwell, Dullege, and Lusum, where, it seems, nobody durst relieve the poor distressed people for fear of the infection.

This notion having, as I said, prevailed with the people in that part of the town, was in part the occasion, as I said before, that they had recourse to ships for their retreat; and where they did this early and with prudence, furnishing themselves so with provisions that they had no need to go on shore for supplies or suffer boats to come on board to bring them—I say, where they did so they had certainly the safest retreat of any people whatsoever; but the distress was such that people ran on board, in their fright, without bread to eat, and some into ships that had no men on board to remove them farther off, or to take the boat and go down the river to buy provisions where it might be done safely, and these often suffered and were infected on board as much as on shore.

As the richer sort got into ships, so the lower rank got into hoys, smacks, lighters, and fishing-boats; and many, especially watermen, lay in their boats; but those made sad work of it, especially the latter, for, going about for provision, and perhaps to get their subsistence, the infection got in among them and made a fearful havoc; many of the watermen died alone in their wherries as they rid at their roads, as well as above bridge as below, and were not found sometimes till they were not in condition for anybody to touch or come near them.

Indeed, the distress of the people at this seafaring end of the town was very deplorable, and deserved the greatest commiseration. But, alas I this was a time when every one's private safety lay so near them that they had no room to pity the distresses of others; for every one had death, as it were, at his door, and many even in their families, and knew not what to do or whither to fly.

This, I say, took away all compassion; self-preservation, indeed, appeared here to be the first law. For the children ran away from their parents as they languished in the utmost distress. And in some places, though not so frequent as the other, parents did the like to their children; nay, some dreadful examples there were, and particularly two in one week, of distressed mothers, raving and distracted, killing their own children; one whereof was not far off from where I dwelt, the poor lunatic creature not living herself long enough to be sensible of the sin of what she had done, much less to be punished for it.

It is not, indeed, to be wondered at: for the danger of immediate death to ourselves took away all bowels of love, all concern for one another. I speak in general, for there were many instances of immovable affection, pity, and duty in many, and some that came to my knowledge, that is to say, by hearsay; for I shall not take upon me to vouch the truth of the particulars.

To introduce one, let me first mention that one of the most deplorable cases in all the present calamity was that of women with child, who, when they came to the hour of their sorrows, and their pains come upon them, could neither have help of one kind or another; neither midwife or neighbouring women to come near them. Most of the midwives were dead, especially of such as served the poor; and many, if not all the midwives of note, were fled into the country; so that it was next to impossible for a poor woman that could not pay an immoderate price to get any midwife to come to her—and if they did, those they could get were generally unskilful and ignorant creatures; and the consequence of this was that a most unusual and incredible number of women were reduced to the utmost distress. Some were delivered and spoiled by the rashness and ignorance of those who pretended to lay them. Children without number were, I might say, murdered by the same but a more justifiable ignorance: pretending they would save the mother, whatever became of the child; and many times both mother and child were lost in the same manner; and especially where the mother had the distemper, there nobody would come near them and both sometimes perished. Sometimes the mother has died of the plague, and the infant, it may be, half born, or born but not parted from the mother. Some died in the very pains of their travail, and not delivered at all; and so many were the cases of this kind that it is hard to judge of them.

Something of it will appear in the unusual numbers which are put into the weekly bills (though I am far from allowing them to be able to give anything of a full account) under the articles of:

Child-bed.
Abortive and Still-born.
Christmas and Infants.

Take the weeks in which the plague was most violent, and compare them with the weeks before the distemper began, even in the same year. For example:—

Child-bed. Abortive. Still-bora.
From January 3 to January 10 7 1 13
" " 10 " 17 8 6 11
" " 17 " 24 9 5 15
" " 24 " 31 3 2 9
" " 31 to February 7 3 3 8
" February 7 " 14 6 2 11
" " 14 "2152 13
" " 21 " 28 2 2 10
" " 28 to March 7 5 110
48 24 100
From August 1 to August 8 25 5 11
" " 8 " 15 23 6 8
" " 15 " 22 28 4 4
" " 22 " 29 40 6 10
" " 29 to September 5 38 2 11
September 5 " 12 39 23 …
" " 12 " 19 42 5 17
" " 19 " 26 42 6 10
" " 26 to October 3 14 4 9
291 61 80

To the disparity of these numbers it is to be considered and allowed for, that according to our usual opinion who were then upon the spot, there were not one-third of the people in the town during the months of August and September as were in the months of January and February. In a word, the usual number that used to die of these three articles, and, as I hear, did die of them the year before, was thus:—

1664. 1665.
Child-bed 189 Child-bed 625
Abortive and still-born 458 Abortive and still-born 617
647 1242

This inequality, I say, is exceedingly augmented when the numbers of people are considered. I pretend not to make any exact calculation of the numbers of people which were at this time in the city, but I shall make a probable conjecture at that part by-and-by. What I have said now is to explain the misery of those poor creatures above; so that it might well be said, as in the Scripture, Woe be to those who are with child, and to those which give suck in that day. For, indeed, it was a woe to them in particular.

I was not conversant in many particular families where these things happened, but the outcries of the miserable were heard afar off. As to those who were with child, we have seen some calculation made; 291 women dead in child-bed in nine weeks, out of one-third part of the number of whom there usually died in that time but eighty-four of the same disaster. Let the reader calculate the proportion.

There is no room to doubt but the misery of those that gave suck was in proportion as great. Our bills of mortahty could give but little light in this, yet some it did. There were several more than usual starved at nurse, but this was nothing. The misery was where they were, first, starved for want of a nurse, the mother dying and all the family and the infants found dead by them, merely for want; and, if I may speak my opinion, I do believe that many hundreds of poor helpless infants perished in this manner. Secondly, not starved, but poisoned by the nurse. Nay, even where the mother has been nurse, and having received the infection, has poisoned, that is, infected the infant with her milk even before they knew they were infected themselves; nay, and the infant has died in such a case before the mother. I cannot but remember to leave this admonition upon record, if ever such another dreadful visitation should happen in this city, that all women that are with child or that give suck should be gone, if they have any possible means, out of the place, because their misery, if infected, will so much exceed all other people's .

I could tell here dismal stories of having infants being found sucking the breasts of their mothers, or nurses, after they have been dead of the plague. Of a mother in the parish where I lived, who, having a child that was not well, sent for an apothecary to view the child; and when he came, as the relation goes, was giving the child suck at her breast, and to all appearance was herself very well; but when the apothecary came close to her he saw the tokens upon that breast with which she was suckling the child. He was surprised enough, to be sure, but, not willing to fright the poor woman too much, he desired she would give the child into his hand; so he takes the child, and going to a cradle in the room, lays it in, and opening its cloths, found the tokens upon the child too, and both died before he could get home to send a preventive medicine to the father of the child, to whom he had told their condition. Whether the child infected the nurse-mother or the mother the child was not certain, but the last most likely. Likewise of a child brought home to the parents from a nurse that had died of the plague, yet the tender mother would not refuse to take in her child, and laid it in her bosom, by which she was infected; and died with the child in her arms dead also.

It would make the hardest heart move at the instances that were frequently found of tender mothers tending and watching with their dear children, and even dying before them, and sometimes taking the distemper from them and dying, when the child for whom the affectionate heart had been sacrificed has got over it and escaped.

The like of a tradesman in East Smithfield, whose wife was big with child of her first child, and fell in labour, having the plague upon her. He could neither get midwife to assist her or nurse to tend her, and two servants which he kept fled both from her. He ran from house to house like one distracted, but could get no help; the utmost he could get was, that a watchman, who attended at an infected house shut up, promised to send a nurse in the morning. The poor man, with his heart broke, went back, assisted his wife what he could, acted the part of the midwife, brought the child dead into the world, and his wife in about an hour died in his arms, where he held her dead body fast till the morning, when the watchman came and brought the nurse as he had promised; and coming up the stairs (for he had left the door open, or only latched), they found the man sitting with his dead wife in his arms, and so overwhelmed with grief that he died in a few hours after without any sign of the infection upon him, but merely sunk under the weight of his grief.

I have heard also of some who, on the death of their relations, have grown stupid with the insupportable sorrow; and of one, in particular, who was so absolutely overcome with the pressure upon his spirits that by degrees his head sank into his body, so between his shoulders that the crown of his head was very little seen above the bone of his shoulders; and by degrees losing both voice and sense, his face, looking forward, lay against his collarbone and could not be kept up any otherwise, unless held up by the hands of other people; and the poor man never came to himself again, but languished near a year in that condition, and died. Nor was he ever once seen to lift up his eyes or to look upon any particular object.

I cannot undertake to give any other than a summary of such passages as these, because it was not possible to come at the particulars, where sometimes the whole families where such things happened were carried off by the distemper. But there were innumerable cases of this kind which presented to the eye and the ear, even in passing along the streets, as I have hinted above. Nor is it easy to give any story of this or that family which there was not divers parallel stories to be met with of the same kind.

But as I am now talking of the time when the plague raged at the easternmost part of the town—how for a long time the people of those parts had flattered themselves that they should escape, and how they were surprised when it came upon them as it did; for, indeed, it came upon them like an armed man when it did come;—I say, this brings me back to the three poor men who wandered from Wapping, not knowing whither to go or what to do, and whom I mentioned before; one a biscuit-baker, one a sailmaker, and the other a joiner, all of Wapping, or there-abouts.

The sleepiness and security of that part, as I have observed, was such that they not only did not shift for themselves as others did, but they boasted of being safe, and of safety being with them; and many people fled out of the city, and out of the infected suburbs, to Wapping, Ratcliff, Limehouse, Poplar, and such Places, as to Places of security; and it is not at all unlikely that their doing this helped to bring the plague that way faster than it might otherwise have come. For though I am much for people flying away and emptying such a town as this upon the first appearance of a like visitation, and that all people who have any possible retreat should make use of it in time and be gone, yet I must say, when all that will fly are gone, those that are left and must stand it should stand stock-still where they are, and not shift from one end of the town or one part of the town to the other; for that is the bane and mischief of the whole, and they carry the plague from house to house in their very clothes.

Wherefore were we ordered to kill all the dogs and cats, but because as they were domestic animals, and are apt to run from house to house and from street to street, so they are capable of carrying the effluvia or infectious streams of bodies infected even in their furs and hair? And therefore it was that, in the beginning of the infection, an order was published by the Lord Mayor, and by the magistrates, according to the advice of the physicians, that all the dogs and cats should be immediately killed, and an officer was appointed for the execution.

It is incredible, if their account is to be depended upon, what a prodigious number of those creatures were destroyed. I think they talked of forty thousand dogs, and five times as many cats; few houses being without a cat, some having several, sometimes five or six in a house. All possible endeavours were used also to destroy the mice and rats, especially the latter, by laying ratsbane and other poisons for them, and a prodigious multitude of them were also destroyed.

I often reflected upon the unprovided condition that the whole body of the people were in at the first coming of this calamity upon them, and how it was for want of timely entering into measures and managements, as well public as private, that all the confusions that followed were brought upon us, and that such a prodigious number of people sank in that disaster, which, if proper steps had been taken, might. Providence concurring, have been avoided, and which, if posterity think fit, they may take a caution and warning from. But I shall come to this part again.

I come back to my three men. Their story has a moral in every part of it, and their whole conduct, and that of some whom they joined with, is a pattern for all poor men to follow, or women either, if ever such a time comes again; and if there was no other end in recording it, I think this a very just one, whether my account be exactly according to fact or no.

Two of them are said to be brothers, the one an old soldier, but now a biscuit-maker; the other a lame sailor, but now a sailmaker; the third a joiner. Says John the biscuit-maker one day to Thomas his brother, the sailmaker, 'Brother Tom, what will become of us? The plague grows hot in the city, and increases this way. What shall we do?'

'Truly,' says Thomas, 'I am at a great loss what to do, for I find if it comes down into Wapping I shall be turned out of my lodging.' And thus they began to talk of it beforehand.

JOHN. Turned out of your lodging, Tom I If you are, I don't know who will take you in; for people are so afraid of one another now, there's no getting a lodging anywhere.

THOMAS. Why, the people where I lodge are good, civil people, and have kindness enough for me too; but they say I go abroad every day to my work, and it will be dangerous; and they talk of locking themselves up and letting nobody come near them.

JOHN. Why, they are in the right, to be sure, if they resolve to venture staying in town.

THOMAS. Nay, I might even resolve to stay within doors too, for, except a suit of sails that my master has in hand, and which I am just finishing, I am like to get no more work a great while. There's no trade stirs now. Workmen and servants are turned off everywhere, so that I might be glad to be locked up too; but I do not see they will be willing to consent to that, any more than to the other.

JOHN. Why, what will you do then, brother? And what shall I do? for I am almost as bad as you. The people where I lodge are all gone into the country but a maid, and she is to go next week, and to shut the house quite up, so that I shall be turned adrift to the wide world before you, and I am resolved to go away too, if I knew but where to go.

THOMAS. We were both distracted we did not go away at first; then we might have travelled anywhere. There's no stirring now; we shall be starved if we pretend to go out of town. They won't let us have victuals, no, not for our money, nor let us come into the towns, much less into their houses.

JOHN. And that which is almost as bad, I have but little money to help myself with neither.

THOMAS. As to that, we might make shift, I have a little, though not much; but I tell you there's no stirring on the road. I know a couple of poor honest men in our street have attempted to travel, and at Barnet, or Whetstone, or thereabouts, the people offered to fire at them if they pretended to go forward, so they are come back again quite discouraged.

JOHN. I would have ventured their fire if I had been there. If I had been denied food for my money they should have seen me take it before their faces, and if I had tendered money for it they could not have taken any course with me by law.

THOMAS. You talk your old soldier's language, as if you were in the Low Countries now, but this is a serious thing. The people have good reason to keep anybody off that they are not satisfied are sound, at such a time as this, and we must not plunder them.

JOHN. No, brother, you mistake the case, and mistake me too. I would plunder nobody; but for any town upon the road to deny me leave to pass through the town in the open highway, and deny me provisions for my money, is to say the town has a right to starve me to death, which cannot be true.

THOMAS. But they do not deny you liberty to go back again from whence you came, and therefore they do not starve you.

JOHN. But the next town behind me will, by the same rule, deny me leave to go back, and so they do starve me between them. Besides, there is no law to prohibit my travelling wherever I will on the road.

THOMAS. But there will be so much difficulty in disputing with them at every town on the road that it is not for poor men to do it or undertake it, at such a time as this is especially.

John. Why, brother, our condition at this rate is worse than anybody else's, for we can neither go away nor stay here. I am of the same mind with the lepers of Samaria: If we stay here we are sure to die', I mean especially as you and I are stated, without a dwelling-house of our own, and without lodging in anybody else's . There is no lying in the street at such a time as this; we had as good go into the dead-cart at once. Therefore I say, if we stay here we are sure to die, and if we go away we can but die; I am resolved to be gone.

THOMAS. You will go away. Whither will you go, and what can you do? I would as willingly go away as you, if I knew whither. But we have no acquaintance, no friends. Here we were born, and here we must die.

JOHN. Look you, Tom, the whole kingdom is my native country as well as this town. You may as well say I must not go out of my house if it is on fire as that I must not go out of the town I was born in when it is infected with the plague. I was bora in England, and have a right to live in it if I can.

THOMAS. But you know every vagrant person may by the laws of England be taken up, and passed back to their last legal settlement.

JOHN. But how shall they make me vagrant? I desire only to travel on, upon my lawful occasions.

THOMAS. What lawful occasions can we pretend to travel, or rather wander upon? They will not be put off with words.

JOHN. Is not flying to save our lives a lawful occasion? And do they not all know that the fact is true? We cannot be said to dissemble.

THOMAS. But suppose they let us pass, whither shall we go?

JOHN. Anywhere, to save our lives; it is time enough to consider that when we are got out of this town. If I am once out of this dreadful place, I care not where I go.

THOMAS. We shall be driven to great extremities. I know not what to think of it.

JOHN. Well, Tom, consider of it a little.

This was about the beginning of July; and though the plague was come forward in the west and north parts of the town, yet all Wapping, as I have observed before, and Redriff, and Ratcliff, and Limehouse, and Poplar, in short, Deptford and Greenwich, all both sides of the river from the Hermitage, and from over against it, quite down to Blackwall, was entirely free; there had not one person died of the plague in all Stepney parish, and not one on the south side of Whitechappel Road, no, not in any parish; and yet the weekly bill was that very week risen up to 1006.

It was a fortnight after this before the two brothers met again, and then the case was a little altered, and the' plague was exceedingly advanced and the number greatly increased; the bill was up at 2785, and prodigiously increasing, though still both sides of the river, as below, kept pretty well. But some began to die in Redriff, and about five or six in Ratcliff Highway, when the sailmaker came to his brother John express, and in some fright; for he was absolutely warned out of his lodging, and had only a week to provide himself. His brother John was in as bad a case, for he was quite out, and had only begged leave of his master, the biscuit-maker, to lodge in an outhouse belonging to his workhouse, where he only lay upon straw, with some biscuit-sacks, or bread-sacks, as they called them, laid upon it, and some of the same sacks to cover him.

Here they resolved (seeing all employment being at an end, and no work or wages to be had), they would make the best of their way to get out of the reach of the dreadful infection, and, being as good husbands as they could, would endeavour to live upon what they had as long as it would last, and then work for more if they could get work anywhere, of any kind, let it be what it would.

While they were considering to put this resolution in practice in the best manner they could, the third man, who was acquainted very well with the sailmaker, came to know of the design, and got leave to be one of the number; and thus they prepared to set out.

It happened that they had not an equal share of money; but as the sailmaker, who had the best stock, was, besides his being lame, the most unfit to expect to get anything by working in the country, so he was content that what money they had should all go into one public stock, on condition that whatever any one of them could gain more than another, it should without any grudging be all added to the public stock.

They resolved to load themselves with as little baggage as possible because they resolved at first to travel on foot, and to go a great way that they might, if possible, be effectually safe; and a great many consultations they had with themselves before they could agree about what way they should travel, which they were so far from adjusting that even to the morning they set out they were not resolved on it.

At last the seaman put in a hint that determined it. 'First,' says he, 'the weather is very hot, and therefore I am for travelling north, that we may not have the sun upon our faces and beating on our breasts, which will heat and suffocate us; and I have been told', says he, 'that it is not good to overheat our blood at a time when, for aught we know, the infection may be in the very air. In the next place,' says he, 'I am for going the way that may be contrary to the wind, as it may blow when we set out, that we may not have the wind blow the air of the city on our backs as we go.' These two cautions were approved of, if it could be brought so to hit that the wind might not be in the south when they set out to go north.

John the baker, who bad been a soldier, then put in his opinion. 'First,' says he, 'we none of us expect to get any lodging on the road, and it will be a little too hard to lie just in the open air. Though it be warm weather, yet it may be wet and damp, and we have a double reason to take care of our healths at such a time as this; and therefore,' says he, 'you, brother Tom, that are a sailmaker, might easily make us a little tent, and I will undertake to set it up every night, and take it down, and a fig for all the inns in England; if we have a good tent over our heads we shall do well enough.'

The joiner opposed this, and told them, let them leave that to him; he would undertake to build them a house every night with his hatchet and mallet, though he had no other tools, which should be fully to their satisfaction, and as good as a tent.

The soldier and the joiner disputed that point some time, but at last the soldier carried it for a tent. The only objection against it was, that it must be carried with them, and that would increase their baggage too much, the weather being hot; but the sailmaker had a piece of good hap fell in which made that easy, for his master whom he worked for, having a rope-walk as well as sailmaking trade, had a little, poor horse that he made no use of then; and being willing to assist the three honest men, he gave them the horse for the carrying their baggage; also for a small matter of three days' work that his man did for him before he went, he let him have an old topgallant sail that was worn out, but was sufficient and more than enough to make a very good tent. The soldier showed how to shape it, and they soon by his direction made their tent, and fitted it with poles or staves for the purpose; and thus they were furnished for their journey, viz., three men, one tent, one horse, one gun—for the soldier would not go without arms, for now he said he was no more a biscuit-baker, but a trooper.

The joiner had a small bag of tools such as might be useful if he should get any work abroad, as well for their subsistence as his own. What money they had they brought all into one public stock, and thus they began their journey. It seems that in the morning when they set out the wind blew, as the sailor said, by his pocket-compass, at N.W. by W. So they directed, or rather resolved to direct, their course N.W.

But then a difficulty came in their way, that, as they set out from the hither end of Wapping, near the Hermitage, and that the plague was now very violent, especially on the north side of the city, as in Shoreditch and Cripplegate parish, they did not think it safe for them to go near those parts; so they went away east through Ratcliff Highway as far as Ratcliff Cross, and leaving Stepney Church still on their left hand, being afraid to come up from Ratcliff Cross to Mile End, because they must come just by the churchyard, and because the wind, that seemed to blow more from the west, blew directly from the side of the city where the plague was hottest. So, I say, leaving Stepney they fetched a long compass, and going to Poplar and Bromley, came into the great road just at Bow.

Here the watch placed upon Bow Bridge would have questioned them, but they, crossing the road into a narrow way that turns out of the hither end of the town of Bow to Old Ford, avoided any inquiry there, and travelled to Old Ford. The constables everywhere were upon their guard not so much. It seems, to stop people passing by as to stop them from taking up their abode in their towns, and withal because of a report that was newly raised at that time: and that, indeed, was not very improbable, viz., that the poor people in London, being distressed and starved for want of work, and by that means for want of bread, were up in arms and had raised a tumult, and that they would come out to all the towns round to plunder for bread. This, I say, was only a rumour, and it was very well it was no more. But it was not so far off from being a reality as it has been thought, for in a few weeks more the poor people became so desperate by the calamity they suffered that they were with great difficulty kept from g out into the fields and towns, and tearing all in pieces wherever they came; and, as I have observed before, nothing hindered them but that the plague raged so violently and fell in upon them so furiously that they rather went to the grave by thousands than into the fields in mobs by thousands; for, in the parts about the parishes of St Sepulcher, Clarkenwell, Cripplegate, Bishopsgate, and Shoreditch, which were the places where the mob began to threaten, the distemper came on so furiously that there died in those few parishes even then, before the plague was come to its height, no less than 5361 people in the first three weeks in August; when at the same time the parts about Wapping, Radcliffe, and Rotherhith were, as before described, hardly touched, or but very lightly; so that in a word though, as I said before, the good management of the Lord Mayor and justices did much to prevent the rage and desperation of the people from breaking out in rabbles and tumults, and in short from the poor plundering the rich—I say, though they did much, the dead-carts did more: for as I have said that in five parishes only there died above 5000 in twenty days, so there might be probably three times that number sick all that time; for some recovered, and great numbers fell sick every day and died afterwards. Besides, I must still be allowed to say that if the bills of mortality said five thousand, I always believed it was near twice as many in reality, there being no room to believe that the account they gave was right, or that indeed they were among such confusions as I saw them in, in any condition to keep an exact account.

But to return to my travellers. Here they were only examined, and as they seemed rather coming from the country than from the city, they found the people the easier with them; that they talked to them, let them come into a public-house where the constable and his warders were, and gave them drink and some victuals which greatly refreshed and encouraged them; and here it came into their heads to say, when they should be inquired of afterwards, not that they came from London, but that they came out of Essex.

To forward this little fraud, they obtained so much favour of the constable at Old Ford as to give them a certificate of their passing from Essex through that village, and that they had not been at London; which, though false in the common acceptance of London in the county, yet was literally true, Wapping or Ratcliff being no part either of the city or hberty.

This certificate directed to the next constable that was at Homerton, one of the hamlets of the parish of Hackney, was so serviceable to them that it procured them, not a free passage there only, but a full certificate of health from a justice of the peace, who upon the constable's application granted it without much difficulty; and thus they passed through the long divided town of Hackney (for it lay then in several separated hamlets), and travelled on till they came into the great north road on the top of Stamford Hill.

By this time they began to be weary, and so in the back-road from Hackney, a little before it opened into the said great road, they resolved to set up their tent and encamp for the first night, which they did accordingly, with this addition, that finding a barn, or a building like a barn, and first searching as well as they could to be sure there was nobody in it, they set up their tent, with the head of it against the barn. This they did also because the wind blew that night very high, and they were but young at such a way of lodging, as well as at the managing their tent.

Here they went to sleep; but the joiner, a grave and sober man, and not pleased with their lying at this loose rate the first night, could not sleep, and resolved, after trying to sleep to no purpose, that he would get out, and, taking the gun in his hand, stand sentinel and guard his companions. So with the gun in his hand, he walked to and again before the barn, for that stood in the field near the road, but within the hedge. He had not been long upon the scout but he heard a noise of people coming on, as if it had been a great number, and they came on, as he thought, directly towards the barn. He did not presently awake his companions; but in a few minutes more, their noise growing louder and louder, the biscuit-baker called to him and asked him what was the matter, and quickly started out too. The other, being the lame sailmaker and most weary, lay still in the tent.

As they expected, so the people whom they had heard came on directly to the barn, when one of our travellers challenged, like soldiers upon the guard, with 'Who comes there?' The people did not answer immediately, but one of them speaking to another that was behind him, 'Alas I alas I we are all disappointed,' says he. 'Here are some people before us; the barn is taken up.'

They all stopped upon that, as under some surprise, and it seems there was about thirteen of them in all, and some women among them. They consulted together what they should do, and by their discourse our travellers soon found they were poor, distressed people too, like themselves, seeking shelter and safety; and besides, our travellers had no need to be afraid of their coming up to disturb them, for as soon as they heard the words, 'Who comes there?' these could hear the women say, as if frighted, 'Do not go near them. How do you know but they may have the plague?' And when one of the men said, 'Let us but speak to them', the women said, 'No, don't by any means. We have escaped thus far by the goodness of God; do not let us run into danger now, we beseech you.'

Our travellers found by this that they were a good, sober sort of people, and flying for their lives, as they were; and, as they were encouraged by it, so John said to the joiner, his comrade, 'Let us encourage them too as much as we can'; so he called to them, 'Hark ye, good people,' says the joiner, 'we find by your talk that you are flying from the same dreadful enemy as we are. Do not be afraid of us; we are only three poor men of us. If you are free from the distemper you shall not be hurt by us. We are not in the barn, but in a little tent here in the outside, and we will remove for you; we can set up our tent again immediately anywhere else'; and upon this a parley began between the joiner, whose name was Richard, and one of their men, who said his name was Ford.

FORD. And do you assure us that you are all sound men?

RICHARD. Nay, we are concerned to tell you of it, that you may not be uneasy or think yourselves in danger; but you see we do not desire you should put yourselves into any danger, and therefore I tell you that we have not made use of the barn, so we will remove from it, that you may be safe and we also.

FORD. That is very kind and charitable; but if we have reason to be satisfied that you are sound and free from the visitation, why should we make you remove now you are settled in your lodging, and, it may be, are laid down to rest? We will go into the barn, if you please, to rest ourselves a while, and we need not disturb you.

RICHARD. Well, but you are more than we are. I hope you will assure us that you are all of you sound too, for the danger is as great from you to us as from us to you.

FORD. Blessed be God that some do escape, though it is but few; what may be our portion still we know not, but hitherto we are preserved.

RICHARD. What part of the town do you come from? Was the plague come to the places where you lived?

FORD. Ay, ay, in a most frightful and terrible manner, or else we had not fled away as we do; but we believe there will be very few left alive behind us.

RICHARD. What part do you come from?

FORD. We are most of us of Cripplegate parish, only two or three of Clerkenwell parish, but on the hither side.

RICHARD. How then was it that you came away no sooner?

FORD. We have been away some time, and kept together as well as we could at the hither end of Islington, where we got leave to lie in an old uninhabited house, and had some bedding and conveniences of our own that we brought with us; but the plague is come up into Islington too, and a house next door to our poor dwelling was infected and shut up; and we are come away in a fright.

RICHARD. And what way are you going?

FORD. As our lot shall cast us; we know not whither, but God will guide those that look up to Him.

    They parleyed no further at that time, but came all up to the barn, and with some difficulty got into it. There was nothing but hay in the barn, but it was almost full of that, and they accommodated themselves as well as they could, and went to rest; but our travellers observed that before they went to sleep an ancient man who it seems was father of one of the women, went to prayer with all the company, recommending themselves to the blessing and direction of Providence, before they went to sleep.

It was soon day at that time of the year, and as Richard the joiner had kept guard the first part of the night, so John the soldier relieved him, and he had the post in the morning, and they began to be acquainted with one another. It seems when they left Islington they intended to have gone north, away to Highgate, but were stopped at Holloway, and there they would not let them pass; so they crossed over the fields and hills to the eastward, and came out at the Boarded River, and so avoiding the towns, they left Hornsey on the left hand and Newington on the right hand, and came into the great road about Stamford Hill on that side, as the three travellers had done on the other side. And now they had thoughts of going over the river in the marshes, and make forwards to Epping Forest, where they hoped they should get leave to rest. It seems they were not poor, at least not so poor as to be in want; at least they had enough to subsist them moderately for two or three months, when, as they said, they were in hopes the cold weather would check the infection, or at least the violence of it would have spent itself, and would abate, if it were only for want of people left alive to he infected.

This was much the fate of our three travellers, only that they seemed to be the better furnished for travelling, and had it in their view to go farther off; for as to the first, they did not propose to go farther than one day's journey, that so they might have intelligence every two or three days how things were at London.

But here our travellers found themselves under an unexpected inconvenience: namely that of their horse, for by means of the horse to carry their baggage they were obliged to keep in the road, whereas the people of this other band went over the fields or roads, path or no path, way or no way, as they pleased; neither had they any occasion to pass through any town, or come near any town, other than to buy such things as they wanted for their necessary subsistence, and in that indeed they were put to much difficulty; of which in its place.

But our three travellers were obliged to keep the road, or else they must commit spoil, and do the country a great deal of damage in breaking down fences and gates to go over enclosed fields, which they were loth to do if they could help it.

Our three travellers, however, had a great mind to join themselves to this company and take their lot with them; and after some discourse they laid aside their first design which looked northward, and resolved to follow the other into Essex; so in the morning they took up their tent and loaded their horse, and away they travelled all together.

They had some difficulty in passing the ferry at the river-side, the ferryman being afraid of them; but after some parley at a distance, the ferryman was content to bring his boat to a place distant from the usual ferry, and leave it there for them to take it; so putting themselves over, he directed them to leave the boat, and he, having another boat, said he would fetch it again, which it seems, however, he did not do for above eight days.

Here, giving the ferryman money beforehand, they had a supply of victuals and drink, which he brought and left in the boat for them; but not without, as I said, having received the money beforehand. But now our travellers were at a great loss and difficulty how to get the horse over, the boat being small and not fit for it: and at last could not do it without unloading the baggage and making him swim over.

From the river they travelled towards the forest, but when they came to Walthamstow the people of that town denied to admit them, as was the case everywhere. The constables and their watchmen kept them off at a distance and parleyed with them. They gave the same account of themselves as before, but these gave no credit to what they said, giving it for a reason that two or three companies had already come that way and made the like pretences, but that they had given several people the distemper in the towns where they had passed; and had been afterwards so hardly used by the country (though with justice, too, as they had deserved) that about Brentwood, or that way, several of them perished in the fields—whether of the plague or of mere want and distress they could not tell.

This was a good reason indeed why the people of Walthamstow should be very cautious, and why they should resolve not to entertain anybody that they were not well satisfied of. But, as Richard the joiner and one of the other men who parleyed with them told them, it was no reason why they should block up the roads and refuse to let people pass through the town, and who asked nothing of them but to go through the street; that if their people were afraid of them, they might go into their houses and shut their doors; they would neither show them civility nor incivility, but go on about their business.

The constables and attendants, not to be persuaded by reason, continued obstinate, and would hearken to nothing; so the two men that talked with them went back to their fellows to consult what was to be done. It was very discouraging in the whole, and they knew not what to do for a good while; but at last John the soldier and biscuit-maker, considering a while, 'Come,' says he, 'leave the rest of the parley to me.' He had not appeared yet, so he sets the joiner, Richard, to work to cut some poles out of the trees and shape them as like guns as he could, and in a little time he had five or six fair muskets, which at a distance would not be known; and about the part where the lock of a gun is he caused them to wrap cloth and rags such as they had, as soldiers do in wet weather to preserve the locks of their pieces from rust; the rest was discoloured with clay or mud, such as they could get; and all this while the rest of them sat under the trees by his direction, in two or three bodies, where they made fires at a good distance from one another.

While this was doing he advanced himself and two or three with him, and set up their tent in the lane within sight of the barrier which the town's men had made, and set a sentinel just by it with the real gun, the only one they had, and who walked to and fro with the gun on his shoulder, so as that the people of the town might see them. Also, he tied the horse to a gate in the hedge just by, and got some dry sticks together and kindled a fire on the other side of the tent, so that the people of the town could see the fire and the smoke, but could not see what they were doing at it.

After the country people had looked upon them very earnestly a great while, and, by all that they could see, could not but suppose that they were a great many in company, they began to be uneasy, not for their going away, but for staying where they were; and above all, perceiving they had horses and arms, for they had seen one horse and one gun at the tent, and they had seen others of them walk about the field on the inside of the hedge by the side of the lane with their muskets, as they took them to be, shouldered; I say, upon such a sight as this, you may be assured they were alarmed and terribly frighted, and it seems they went to a justice of the peace to know what they should do. What the justice advised them to I know not, but towards the evening they called from the barrier, as above, to the sentinel at the tent.

'What do you want?' says John.*

[* It seems John was in the tent, but hearing them call, he steps out, and taking the gun upon his shoulder, talked to them as if he had been the sentinel placed there upon the guard by some officer that was his superior.]

'Why, what do you intend to do?' says the constable. 'To do,' says John; 'what would you have us to do?'

CONSTABLE. Why don't you be gone? What do you stay there for?

JOHN. Why do you stop us on the king's highway, and pretend to refuse us leave to go on our way?

CONSTABLE. We are not bound to tell you our reason, though we did let you know it was because of the plague.

JOHN. We told you we were all sound and free from the plague, which we were not bound to have satisfied you of, and yet you pretend to stop us on the highway.

CONSTABLE. We have a right to stop it up, and our own safety obliges us to it. Besides, this is not the king's highway; 'tis a way upon sufferance. You see here is a gate, and if we do let people pass here, we make them pay toll.

JOHN. We have a right to seek our own safety as well as you, and you may see we are flying for our lives: and 'tis very unchristian and unjust to stop us.

CONSTABLE. You may go back from whence you came; we do not hinder you from that.

JOHN. No; it is a stronger enemy than you that keeps us from doing that, or else we should not have come hither.

CONSTABLE. Well, you may go any other way, then.

JOHN. No, no; I suppose you see we are able to send you going, and all the people of your parish, and come through your town when we will; but since you have stopped us here, we are content.

You see we have encamped here, and here we will live. We hope you will furnish us with victuals.

Constable. We furnish you? What mean you by that?

John. Why, you would not have us starve, would you? If you stop us here, you must keep us.

Constable. You will be ill kept at our maintenance.

John. If you stint us, we shall make ourselves the better allowance.

Constable. Why, you will not pretend to quarter upon us by force, will you?

John. We have offered no violence to you yet. Why do you seem to oblige us to it? I am an old soldier, and cannot starve, and if you think that we shall be obliged to go back for want of provisions, you are mistaken.

Constable. Since you threaten us, we shall take care to be strong enough for you. I have orders to raise the county upon you.

John. It is you that threaten, not we. And since you are for mischief, you cannot blame us if we do not give you time for it; we shall begin our march in a few minutes.*

[* This frighted the constable and the people that were with him, that they immediately changed their note.]

Constable. What is it you demand of us?

JOHN. At first we desired nothing of you but leave to go through the town; we should have offered no injury to any of you, neither would you have had any injury or loss by us. We are not thieves, but poor people in distress, and flying from the dreadful plague in London, which devours thousands every week. We wonder how you could be so unmerciful!

CONSTABLE. Self-preservation obliges us.

JOHN. What! To shut up your compassion in a case of such distress as this?

CONSTABLE. Well, if you will pass over the fields on your left hand, and behind that part of the town, I will endeavour to have gates opened for you.

JOHN. Our horsemen* cannot pass with our baggage that way; it does not lead into the road that we want to go, and why should you force us out of the road? Besides, you have kept us here all day without any provisions but such as we brought with us. I think you ought to send us some provisions for our relief.

[* They had but one horse among them.]

CONSTABLE. If you will go another way we will send you some provisions.

JOHN. That is the way to have all the towns in the county stop up the ways against us.

CONSTABLE. If they all furnish you with food, what will you be the worse? I see you have tents; you want no lodging.

JOHN. Well, what quantity of provisions will you send us?
Constable. How many are you?

JOHN. Nay, we do not ask enough for all our company; we are in three companies. If you will send us bread for twenty men and about six or seven women for three days, and show us the way over the field you speak of, we desire not to put your people into any fear for us; we will go out of our way to oblige you, though we are as free from infection as you are.*

[* Here he called to one of his men, and bade him order Captain Richard and his people to march the lower way on the side of the marches, and meet them in the forest; which was all a sham, for they had no Captain Richard, or any such company.]

CONSTABLE. And will you assure us that your other people shall offer us no new disturbance?

JOHN. No, no you may depend on it.

CONSTABLE. You must oblige yourself, too, that none of your people shall come a step nearer than where the provisions we send you shall be set down.

JOHN. I answer for it we will not.

Accordingly they sent to the place twenty loaves of bread and three or four large pieces of good beef, and opened some gates, through which they passed; but none of them had courage so much as to look out to see them go, and, as it was evening, if they had looked they could not have seen them as to know how few they were.

This was John the soldier's management. But this gave such an alarm to the county, that had they really been two or three hundred the whole county would have been raised upon them, and they would have been sent to prison, or perhaps knocked on the head.

They were soon made sensible of this, for two days afterwards they found several parties of horsemen and footmen also about, in pursuit of three companies of men, armed, as they said, with muskets, who were broke out from London and had the plague upon them, and that were not only spreading the distemper among the people, but plundering the country.

As they saw now the consequence of their case, they soon saw the danger they were in; so they resolved by the advice also of the old soldier to divide themselves again. John and his two comrades, with the horse, went away, as if towards Waltham; the other in two companies, but all a little asunder, and went towards Epping.

The first night they encamped all in the forest, and not far off of one another, but not setting up the tent, lest that should discover them. On the other hand, Richard went to work with his axe and his hatchet, and cutting down branches of trees, he built three tents or hovels, in which they all encamped with as much convenience as they could expect.

The provisions they had at Walthamstow served them very plentifully this night; and as for the next, they left it to Providence. They had fared so well with the old soldier's conduct that they now willingly made him their leader, and the first of his conduct appeared to be very good. He told them that they were now at a proper distance enough from London; that as they need not be immediately beholden to the country for relief, so they ought to be as careful the country did not infect them as that they did not infect the country; that what little money they had, they must be as frugal of as they could; that as he would not have them think of offering the country any violence, so they must endeavour to make the sense of their condition go as far with the country as it could. They all referred themselves to his direction, so they left their three houses standing, and the next day went away towards Epping. The captain also (for so they now called him), and his two fellow-travellers, laid aside their design of going to Waltham, and all went together.

When they came near Epping they halted, choosing out a proper place in the open forest, not very near the highway, but not far out of it on the north side, under a little cluster of low pollard-trees. Here they pitched their little camp—which consisted of three large tents or huts made of poles which their carpenter, and such as were his assistants, cut down and fixed in the ground in a circle, binding all the small ends together at the top and thickening the sides with boughs of trees and bushes, so that they were completely close and warm. They had, besides this, a little tent where the women lay by themselves, and a hut to put the horse in.

It happened that the next day, or next but one, was market-day at Epping, when Captain John and one of the other men went to market and bought some provisions; that is to say, bread, and some mutton and beef; and two of the women went separately, as if they had not belonged to the rest, and bought more. John took the horse to bring it home, and the sack which the carpenter carried his tools in, to put it in. The carpenter went to work and made them benches and stools to sit on, such as the wood he could get would afford, and a kind of table to dine on.

They were taken no notice of for two or three days, but after that abundance of people ran out of the town to look at them, and all the country was alarmed about them. The people at first seemed afraid to come near them; and, on the other hand, they desired the people to keep off, for there was a rumour that the plague was at Waltham, and that it had been in Epping two or three days; so John called out to them not to come to them, 'for,' says he, 'we are all whole and sound people here, and we would not have you bring the plague among us, nor pretend we brought it among you.'

After this the parish officers came up to them and parleyed with them at a distance, and desired to know who they were, and by what authority they pretended to fix their stand at that place. John answered very frankly, they were poor distressed people from London who, foreseeing the misery they should be reduced to if plague spread into the city, had fled out in time for their lives, and, having no acquaintance or relations to fly to, had first taken up at Islington; but, the plague being come into that town, were fled farther; and as they supposed that the people of Epping might have refused them coming into their town, they had pitched their tents thus in the open field and in the forest, being willing to bear all the hardships of such a disconsolate lodging rather than have any one think or be afraid that they should receive injury by them.

At first the Epping people talked roughly to them, and told them they must remove; that this was no place for them; and that they pretended to be sound and well, but that they might be infected with the plague for aught they knew, and might infect the whole country, and they could not suffer them there.

John argued very calmly with them a great while, and told them that London was the place by which they—that is, the townsmen of Epping and all the country round them—subsisted; to whom they sold the produce of their lands, and out of whom they made their rent of their farms; and to be so cruel to the inhabitants of London, or to any of those by whom they gained so much, was very hard, and they would be loth to have it remembered hereafter, and have it told how barbarous, how inhospitable, and how unkind they were to the people of London when they fled from the face of the most terrible enemy in the world; that it would be enough to make the name of an Epping man hateful through all the city, and to have the rabble stone them in the very streets whenever they came so much as to market; that they were not yet secure from being visited themselves, and that, as he heard, Waltham was already; that they would think it very hard that when any of them fled for fear before they were touched, they should be denied the liberty of lying so much as in the open fields.

The Epping men told them again, that they, indeed, said they were sound and free from the infection, but that they had no assurance of it; and that it was reported that there had been a great rabble of people at Walthamstow, who made such pretences of being sound as they did, but that they threatened to plunder the town and force their way, whether the parish officers would or no; that there were near two hundred of them, and had arms and tents like Low Country soldiers; that they extorted provisions from the town, by threatening them with living upon them at free quarter, showing their arms, and talking in the language of soldiers; and that several of them being gone away toward Rumford and Brentwood, the country had been infected by them, and the plague spread into both those large towns, so that the people durst not go to market there as usual; that it was very likely they were some of that party; and if so, they deserved to be sent to the county jail, and be secured till they had made satisfaction for the damage they had done, and for the terror and fright they had put the country into.

John answered that what other people had done was nothing to them; that they assured them they were all of one company; that they had never been more in number than they saw them at that time (which, by the way, was very true); that they came out in two separate companies, but joined by the way, their cases being the same; that they were ready to give what account of themselves anybody could desire of them, and to give in their names and places of abode, that so they might be called to an account for any disorder that they might be guilty of; that the townsmen might see they were content to live hardly, and only desired a little room to breathe in on the forest where it was wholesome; for where it was not they could not stay, and would decamp if they found it otherwise there.

'But,' said the townsmen, 'we have a great charge of poor upon our hands already, and we must take care not to increase it; we suppose you can give us no security against your being chargeable to our parish and to the inhabitants, any more than you can of being dangerous to us as to the infection.'

'Why, look you,' says John, 'as to being chargeable to you, we hope we shall not. If you will relieve us with provisions for our present necessity, we will be very thankful; as we all lived without charity when we were at home, so we will oblige ourselves fully to repay you, if God pleases to bring us back to our own families and houses in safety, and to restore health to the people of London.

'As to our dying here: we assure you, if any of us die, we that survive will bury them, and put you to no expense, except it should be that we should all die; and then, indeed, the last man not being able to bury himself, would put you to that single expense which I am persuaded', says John, 'he would leave enough behind him to pay you for the expense of.

'On the other hand,' says John, 'if you shut up all bowels of compassion, and not relieve us at all, we shall not extort anything by violence or steal from any one; but when what little we have is spent, if we perish for want, God's will be done.'

John wrought so upon the townsmen, by talking thus rationally and smoothly to them, that they went away; and though they did not give any consent to their staying there, yet they did not molest them; and the poor people continued there three or four days longer without any disturbance.

In this time they had got some remote acquaintance with a victualling-house at the outskirts of the town, to whom they called at a distance to bring some little things that they wanted, and which they caused to be set down at a distance, and always paid for very honestly.

During this time the younger people of the town came frequently pretty near them, and would stand and look at them, and sometimes talk with them at some space between; and particularly it was observed that the first Sabbath-day the poor people kept retired, worshipped God together, and were heard to sing psalms.

These things, and a quiet, inoffensive behaviour, began to get them the good opinion of the country, and people began to pity them and speak very well of them; the consequence of which was, that upon the occasion of a very wet, rainy night, a certain gentleman who lived in the neighbourhood sent them a little cart with twelve trusses or bundles of straw, as well for them to lodge upon as to cover and thatch their huts and to keep them dry. The minister of a parish not far off, not knowing of the other, sent them also about two bushels of wheat and half a bushel of white peas.

They were very thankful, to be sure, for this relief, and particularly the straw was a very great comfort to them; for though the ingenious carpenter had made frames for them to lie in like troughs, and filled them with leaves of trees, and such things as they could get, and had cut all their tent-cloth out to make them coverlids, yet they lay damp and hard and unwholesome till this straw came, which was to them like feather-beds, and, as John said, more welcome than feather-beds would have been at another time.

This gentleman and the minister having thus begun, and given an example of charity to these wanderers, others quickly followed, and they received every day some benevolence or other from the people, but chiefly from the gentlemen who dwelt in the country round them. Some sent them chairs, stools, tables, and such household things as they gave notice they wanted; some sent them blankets, rugs, and coverlids, some earthenware, and some kitchen ware for ordering their food.

Encouraged by this good usage, their carpenter in a few days built them a large shed or house with rafters, and a roof in form, and an upper floor, in which they lodged warm: for the weather began to be damp and cold in the beginning of September. But this house, being well thatched, and the sides and roof made very thick, kept out the cold well enough. He made, also, an earthen wall at one end with a chimney in it, and another of the company, with a vast deal of trouble and pains, made a funnel to the chimney to carry out the smoke.

Here they lived comfortably, though coarsely, till the beginning of September, when they had the bad news to hear, whether true or not, that the plague, which was very hot at Waltham Abbey on one side and at Rumford and Brentwood on the other side, was also coming to Epping, to Woodford, and to most of the towns upon the Forest, and which, as they said, was brought down among them chiefly by the higlers, and such people as went to and from London with provisions.

If this was true, it was an evident contradiction to that report which was afterwards spread all over England, but which, as I have said, I cannot confirm of my own knowledge: namely, that the market-people carrying provisions to the city never got the infection or carried it back into the country; both which, I have been assured, has been false.

It might be that they were preserved even beyond expectation, though not to a miracle, that abundance went and came and were not touched; and that was much for the encouragement of the poor people of London, who had been completely miserable if the people that brought provisions to the markets had not been many times wonderfully preserved, or at least more preserved than could be reasonably expected.

But now these new inmates began to be disturbed more effectually, for the towns about them were really infected, and they began to be afraid to trust one another so much as to go abroad for such things as they wanted, and this pinched them very hard, for now they had little or nothing but what the charitable gentlemen of the country supplied them with. But, for their encouragement, it happened that other gentlemen in the country who had not sent them anything before, began to hear of them and supply them, and one sent them a large pig—that is to say, a porker another two sheep, and another sent them a calf. In short, they had meat enough, and sometimes had cheese and milk, and all such things. They were chiefly put to it for bread, for when the gentlemen sent them corn they had nowhere to bake it or to grind it. This made them eat the first two bushel of wheat that was sent them in parched corn, as the Israelites of old did, without grinding or making bread of it.

At last they found means to carry their com to a windmill near Woodford, where they bad it ground, and afterwards the biscuit-maker made a hearth so hollow and dry that he could bake biscuit-cakes tolerably well; and thus they came into a condition to live without any assistance or supplies from the towns; and it was well they did, for the country was soon after fully infected, and about 120 were said to have died of the distemper in the villages near them, which was a terrible thing to them.

On this they called a new council, and now the towns had no need to be afraid they should settle near them; but, on the contrary, several families of the poorer sort of the inhabitants quitted their houses and built huts in the forest after the same manner as they had done. But it was observed that several of these poor people that had so removed had the sickness even in their huts or booths; the reason of which was plain, namely, not because they removed into the air, but, (1) because they did not remove time enough; that is to say, not till, by openly conversing with the other people their neighbours, they had the distemper upon them, or (as may be said) among them, and so carried it about them whither they went. Or (2) because they were not careful enough, after they were safely removed out of the towns, not to come in again and mingle with the diseased people.

But be it which of these it will, when our travellers began to perceive that the plague was not only in the towns, but even in the tents and huts on the forest near them, they began then not only to be afraid, but to think of decamping and removing; for had they stayed they would have been in manifest danger of their lives.

It is not to be wondered that they were greatly afflicted at being obliged to quit the place where they had been so kindly received, and where they had been treated with so much humanity and charity; but necessity and the hazard of life, which they came out so far to preserve, prevailed with them, and they saw no remedy. John, however, thought of a remedy for their present misfortune: namely, that he would first acquaint that gentleman who was their principal benefactor with the distress they were in, and to crave his assistance and advice.

The good, charitable gentleman encouraged them to quit the Place for fear they should be cut off from any retreat at all by the violence of the distemper; but whither they should go, that he found very hard to direct them to. At last John asked of him whether he, being a justice of the peace, would give them certificates of health to other justices whom they might come before; that so whatever might be their lot, they might not be repulsed now they had been also so long from London. This his worship immediately granted, and gave them proper letters of health, and from thence they were at liberty to travel whither they pleased.

Accordingly they had a full certificate of health, intimating that they had resided in a village in the county of Essex so long that, being examined and scrutinised sufficiently, and having been retired from all conversation for above forty days, without any appearance of sickness, they were therefore certainly concluded to be sound men, and might be safely entertained anywhere, having at last removed rather for fear of the plague which was come into such a town, rather than for having any signal of infection upon them, or upon any belonging to them.

With this certificate they removed, though with great reluctance; and John inclining not to go far from home, they moved towards the marshes on the side of Waltham. But here they found a man who, it seems, kept a weir or stop upon the river, made to raise the water for the barges which go up and down the river, and he terrified them with dismal stories of the sickness having been spread into all the towns on the river and near the river, on the side of Middlesex and Hertfordshire; that is to say, into Waltham, Waltham Cross, Enfield, and Ware, and all the towns on the road, that they were afraid to go that way; though it seems the man imposed upon them, for that the thing was not really true.

However, it terrified them, and they resolved to move across the forest towards Rumford and Brentwood; but they heard that there were numbers of people fled out of London that way, who lay up and down in the forest called Henalt Forest, reaching near Rumford, and who, having no subsistence or habitation, not only lived oddly and suffered great extremities in the woods and fields for want of relief, but were said to be made so desperate by those extremities as that they offered many violences to the county robbed and plundered, and killed cattle, and the like; that others, building huts and hovels by the roadside, begged, and that with an importunity next door to demanding relief; so that the county was very uneasy, and had been obliged to take some of them up.

This in the first place intimated to them, that they would be sure to find the charity and kindness of the county, which they had found here where they were before, hardened and shut up against them; and that, on the other hand, they would be questioned wherever they came, and would be in danger of violence from others in like cases as themselves.

Upon all these considerations John, their captain, in all their names, went back to their good friend and benefactor, who had relieved them before, and laying their case truly before him, humbly asked his advice; and he as kindly advised them to take up their old quarters again, or if not, to remove but a little farther out of the road, and directed them to a proper place for them; and as they really wanted some house rather than huts to shelter them at that time of the year, it growing on towards Michaelmas, they found an old decayed house which had been formerly some cottage or little habitation but was so out of repair as scarce habitable; and by the consent of a farmer to whose farm it belonged, they got leave to make what use of it they could.

The ingenious joiner, and all the rest, by his directions went to work with it, and in a very few days made it capable to shelter them all in case of bad weather; and in which there was an old chimney and old oven, though both lying in ruins; yet they made them both fit for use, and, raising additions, sheds, and leantos on every side, they soon made the house capable to hold them all.

They chiefly wanted boards to make window-shutters, floors, doors, and several other things; but as the gentlemen above favoured them, and the country was by that means made easy with them, and above all, that they were known to be all sound and in good health, everybody helped them with what they could spare.

Here they encamped for good and all, and resolved to remove no more.

They saw plainly how terribly alarmed that county was everywhere at anybody that came from London, and that they should have no admittance anywhere but with the utmost difficulty; at least no friendly reception and assistance as they had received here.

Now, although they received great assistance and encouragement from the country gentlemen and from the people round about them, yet they were put to great straits: for the weather grew cold and wet in October and November, and they had not been used to so much hardship; so that they got colds in their limbs, and distempers, but never had the infection; and thus about December they came home to the city again.

I give this story thus at large, principally to give an account what became of the great numbers of people which immediately appeared in the city as soon as the sickness abated; for, as I have said, great numbers of those that were able and had retreats in the country fled to those retreats. So, when it was increased to such a frightful extremity as I have related, the middling people who had not friends fled to all parts of the country where they could get shelter, as well those that had money to relieve themselves as those that had not. Those that had money always fled farthest, because they were able to subsist themselves; but those who were empty suffered, as I have said, great hardships, and were often driven by necessity to relieve their wants at the expense of the country. By that means the country was made very uneasy at them, and sometimes took them up; though even then they scarce knew what to do with them, and were always very backward to punish them, but often, too, they forced them from place to place till they were obliged to come back again to London.

I have, since my knowing this story of John and his brother, inquired and found that there were a great many of the poor disconsolate people, as above, fled into the country every way; and some of them got little sheds and barns and outhouses to live in, where they could obtain so much kindness of the country, and especially where they had any the least satisfactory account to give of themselves, and particularly that they did not come out of London too late. But others, and that in great numbers, built themselves little huts and retreats in the fields and woods, and lived like hermits in holes and caves, or any place they could find, and where, we may be sure, they suffered great extremities, such that many of them were obliged to come back again whatever the danger was; and so those little huts were often found empty, and the country people supposed the inhabitants lay dead in them of the plague, and would not go near them for fear—no, not in a great while; nor is it unlikely but that some of the unhappy wanderers might die so all alone, even sometimes for want of help, as particularly in one tent or hut was found a man dead, and on the gate of a field just by was cut with his knife in uneven letters the following words, by which it may be supposed the other man escaped, or that, one dying first, the other buried him as well as he could:—

0 mIsErY!
We **Both** ShaLL DyE,
WoE, WoE.

I have given an account already of what I found to have been the case down the river among the seafaring men; how the ships lay in the offing, as it's called, in rows or lines astern of one an other, quite down from the Pool as far as I could see. I have been told that they lay in the same manner quite down the river as low as Gravesend, and some far beyond: even everywhere or in every place where they could ride with safety as to wind and weather; nor did I ever hear that the plague reached to any of the people on board those ships—except such as lay up in the Pool, or as high as Deptford Reach, although the people went frequently on shore to the country towns and villages and farmers' houses, to buy fresh provisions, fowls, pigs, calves, and the like for their supply.

Likewise I found that the watermen on the river above the bridge found means to convey themselves away up the river as far as they could go, and that they had, many of them, their whole families in their boats, covered with tilts and bales, as they call them, and furnished with straw within for their lodging, and that they lay thus all along by the shore in the marshes, some of them setting up little tents with their sails, and so lying under them on shore in the day, and going into their boats at night; and in this manner, as I have heard, the river-sides were lined with boats and people as long as they had anything to subsist on, or could get anything of the country; and indeed the country people, as well Gentlemen as others, on these and all other occasions, were very forward to relieve them—but they were by no means willing to receive them into their towns and houses, and for that we cannot blame them.

There was one unhappy citizen within my knowledge who had been visited in a dreadful manner, so that his wife and all his children were dead, and himself and two servants only left, with an elderly woman, a near relation, who had nursed those that were dead as well as she could. This disconsolate man goes to a village near the town, though not within the bills of mortality, and finding an empty house there, inquires out the owner, and took the house. After a few days he got a cart and loaded it with goods, and carries them down to the house; the people of the village opposed his driving the cart along; but with some arguings and some force, the men that drove the cart along got through the street up to the door of the house. There the constable resisted them again, and would not let them be brought in. The man caused the goods to be unloaden and laid at the door, and sent the cart away; upon which they carried the man before a justice of peace; that is to say, they commanded him to go, which he did. The justice ordered him to cause the cart to fetch away the goods again, which he refused to do; upon which the justice ordered the constable to pursue the carters and fetch them back, and make them reload the goods and carry them away, or to set them in the stocks till they came for further orders; and if they could not find them, nor the man would not consent to take them away, they should cause them to be drawn with hooks from the house-door and burned in the street. The poor distressed man upon this fetched the goods again, but with grievous cries and lamentations at the hardship of his case. But there was no remedy; self-preservation obliged the people to those severities which they would not otherwise have been concerned in. Whether this poor man lived or died I cannot tell, but it was reported that he had the plague upon him at that time; and perhaps the people might report that to justify their usage of him; but it was not unlikely that either he or his goods, or both, were dangerous, when his whole family had been dead of the distempers so little a while before.

I know that the inhabitants of the towns adjacent to London were much blamed for cruelty to the poor people that ran from the contagion in their distress, and many very severe things were done, as may be seen from what has been said; but I cannot but say also that, where there was room for charity and assistance to the people, without apparent danger to themselves, they were willing enough to help and relieve them. But as every town were indeed judges in their own case, so the poor people who ran abroad in their extremities were often ill-used and driven back again into the town; and this caused infinite exclamations and outcries against the country towns, and made the clamour very popular.

And yet, more or less, maugre all the caution, there was not a town of any note within ten (or, I believe, twenty) miles of the city but what was more or less infected and had some died among them. I have heard the accounts of several, such as they were reckoned up, as follows:—

In Enfield 32 In Uxbridge 117
" Hornsey 58 " Hertford 90
" Newington 17 " Ware 160
" Tottenham 42 " Hodsdon 30
" Edmonton 19 " Waltham Abbey 23
" Barnet and Hadly 19 " Epping 26
" St Albans 121 " Deptford 623
" Watford 45 " Greenwich 231
" Eltham and Lusum 85 " Kingston 122
" Croydon 61 " Stanes 82
" Brentwood 70 " Chertsey 18
" Rumford 109 " Windsor 103
" Barking Abbot 200
" Brentford 432 Cum aliis.

Another thing might render the country more strict with respect to the citizens, and especially with respect to the poor, and this was what I hinted at before: namely, that there was a seeming propensity or a wicked inclination in those that were infected to infect others.

There have been great debates among our physicians as to the reason of this. Some will have it to be in the nature of the disease, and that it impresses every one that is seized upon by it with a kind of a rage, and a hatred against their own kind—as if there was a malignity not only in the distemper to communicate itself, but in the very nature of man, prompting him with evil will or an evil eye, that, as they say in the case of a mad dog, who though the gentlest creature before of any of his kind, yet then will fly upon and bite any one that comes next him, and those as soon as any who had been most observed by him before.

Others placed it to the account of the corruption of human nature, who cannot bear to see itself more miserable than others of its own species, and has a kind of involuntary wish that all men were as unhappy or in as bad a condition as itself.

Others say it was only a kind of desperation, not knowing or regarding what they did, and consequently unconcerned at the danger or safety not only of anybody near them, but even of themselves also. And indeed, when men are once come to a condition to abandon themselves, and be unconcerned for the safety or at the danger of themselves, it cannot be so much wondered that they should be careless of the safety of other people.

But I choose to give this grave debate a quite different turn, and answer it or resolve it all by saying that I do not grant the fact. On the contrary, I say that the thing is not really so, but that it was a general complaint raised by the people inhabiting the outlying villages against the citizens to justify, or at least excuse, those hardships and severities so much talked of, and in which complaints both sides may be said to have injured one another; that is to say, the citizens pressing to be received and harboured in time of distress, and with the plague upon them, complain of the cruelty and injustice of the country people in being refused entrance and forced back again with their goods and families; and the inhabitants, finding themselves so imposed upon, and the citizens breaking in as it were upon them whether they would or no, complain that when they were infected they were not only regardless of others, but even willing to infect them; neither of which were really true—that is to say, in the colours they were described in.

It is true there is something to be said for the frequent alarms which were given to the country of the resolution of the people of London to come out by force, not only for relief, but to plunder and rob; that they ran about the streets with the distemper upon them without any control; and that no care was taken to shut up houses, and confine the sick people from infecting others; whereas, to do the Londoners justice, they never practised such things, except in such particular cases as I have mentioned above, and such like. On the other hand, everything was managed with so much care, and such excellent order was observed in the whole city and suburbs by the care of the Lord Mayor and aldermen and by the justices of the peace, churchwardens, &c., in the outparts, that London may be a pattern to all the cities in the world for the good government and the excellent order that was everywhere kept, even in the time of the most violent infection, and when the people were in the utmost consternation and distress. But of this I shall speak by itself.

One thing, it is to be observed, was owing principally to the prudence of the magistrates, and ought to be mentioned to their honour: viz., the moderation which they used in the great and difficult work of shutting up of houses. It is true, as I have mentioned, that the shutting up of houses was a great subject of discontent, and I may say indeed the only subject of discontent among the people at that time; for the confining the sound in the same house with the sick was counted very terrible, and the complaints of people so confined were very grievous. They were heard into the very streets, and they were sometimes such that called for resentment, though oftener for compassion. They had no way to converse with any of their friends but out at their windows, where they would make such piteous lamentations as often moved the hearts of those they talked with, and of others who, passing by, heard their story; and as those complaints oftentimes reproached the severity, and sometimes the insolence, of the watchmen placed at their doors, those watchmen would answer saucily enough, and perhaps be apt to affront the people who were in the street talking to the said families; for which, or for their ill-treatment of the families, I think seven or eight of them in several places were killed; I know not whether I should say murdered or not, because I cannot enter into the particular cases. It is true the watchmen were on their duty, and acting in the post where they were placed by a lawful authority; and killing any public legal officer in the execution of his office is always, in the language of the law, called murder. But as they were not authorised by the magistrates' instructions, or by the power they acted under, to be injurious or abusive either to the people who were under their observation or to any that concerned themselves for them; so when they did so, they might he said to act themselves, not their office;' to act as private persons, not as persons employed; and consequently, if they brought mischief upon themselves by such an undue behaviour, that mischief was upon their own heads; and indeed they had so much the hearty curses of the people, whether they deserved it or not, that whatever befell them nobody pitied them, and everybody was apt to say they deserved it, whatever it was. Nor do I remember that anybody was ever punished, at least to any considerable degree, for whatever was done to the watchmen that guarded their houses.

What variety of stratagems were used to escape and get out of houses thus shut up, by which the watchmen were deceived or overpowered, and that the people got away, I have taken notice of already, and shall say no more to that. But I say the magistrates did moderate and ease families upon many occasions in this case, and particularly in that of taking away, or suffering to be removed, the sick persons out of such houses when they were willing to be removed either to a pest-house or other Places; and sometimes giving the well persons in the family so shut up, leave to remove upon information given that they were well, and that they would confine themselves in such houses where they went so long as should be required of them. The concern, also, of the magistrates for the supplying such poor families as were infected—I say, supplying them with necessaries, as well physic as food—was very great, and in which they did not content themselves with giving the necessary orders to the officers appointed, but the aldermen in person, and on horseback, frequently rode to such houses and caused the people to be asked at their windows whether they were duly attended or not; also, whether they wanted anything that was necessary, and if the officers had constantly carried their messages and fetched them such things as they wanted or not. And if they answered in the affirmative, all was well; but if they complained that they were ill supplied, and that the officer did not do his duty, or did not treat them civilly, they (the officers) were generally removed, and others placed in their stead.

It is true such complaint might be unjust, and if the officer had such arguments to use as would convince the magistrate that he was right, and that the people had injured him, he was continued and they reproved. But this part could not well bear a particular inquiry, for the parties could very ill be well heard and answered in the street from the windows, as was the case then. The magistrates, therefore, generally chose to favour the people and remove the man, as what seemed to be the least wrong and of the least ill consequence; seeing if the watchman was injured, yet they could easily make him amends by giving him another post of the like nature; but if the family was injured, there was no satisfaction could be made to them, the damage perhaps being irreparable, as it concerned their lives.

A great variety of these cases frequently happened between the watchmen and the poor people shut up, besides those I formerly mentioned about escaping. Sometimes the watchmen were absent, sometimes drunk, sometimes asleep when the people wanted them, and such never failed to be punished severely, as indeed they deserved.

But after all that was or could be done in these cases, the shutting up of houses, so as to confine those that were well with those that were sick, had very great inconveniences in it, and some that were very tragical, and which merited to have been considered if there had been room for it. But it was authorised by a law, it had the public good in view as the end chiefly aimed at, and all the private injuries that were done by the putting it in execution must be put to the account of the public benefit.

It is doubtful to this day whether, in the whole, it contributed anything to the stop of the infection; and indeed I cannot say it did, for nothing could run with greater fury and rage than the infection did when it was in its chief violence, though the houses infected were shut up as exactly and as effectually as it was possible. Certain it is that if all the infected persons were effectually shut in, no sound person could have been infected by them, because they could not have come near them. But the case was this (and I shall only touch it here): namely, that the infection was propagated insensibly, and by such persons as were not visibly infected, who neither knew whom they infected or who they were infected by.

A house in Whitechappel was shut up for the sake of one infected maid, who had only spots, not the tokens come out upon her, and recovered; yet these people obtained no liberty to stir, neither for air or exercise, forty days. Want of breath, fear, anger, vexation, and all the other gifts attending such an injurious treatment cast the mistress of the family into a fever, and visitors came into the house and said it was the plague, though the physicians declared it was not. However, the family were obliged to begin their quarantine anew on the report of the visitors or examiner, though their former quarantine wanted but a few days of being finished. This oppressed them so with anger and grief, and, as before, straitened them also so much as to room, and for want of breathing and free air, that most of the family fell sick, one of one distemper, one of another, chiefly scorbutic ailments; only one, a violent colic; till, after several prolongings of their confinement, some or other of those that came in with the visitors to inspect the persons that were ill, in hopes of releasing them, brought the distemper with them and infected the whole house; and all or most of them died, not of the plague as really upon them before, but of the plague that those people brought them, who should have been careful to have protected them from it. And this was a thing which frequently happened, and was indeed one of the worst consequences of shutting houses up.

I had about this time a little hardship put upon me, which I was at first greatly afflicted at, and very much disturbed about though, as it proved, it did not expose me to any disaster; and this was being appointed by the alderman of Portsoken Ward one of the examiners of the houses in the precinct where I lived. We had a large parish, and had no less than eighteen examiners, as the order called us; the people called us visitors. I endeavoured with all my might to be excused from such an employment, and used many arguments with the alderman's deputy to be excused; particularly I alleged that I was against shutting up houses at all, and that it would be very hard to oblige me to be an instrument in that which was against my judgement, and which I did verily behve would not answer the end it was intended for; but all the abatement I could get was only, that whereas the officer was appointed by my Lord Mayor to continue two months, I should be obliged to hold it but three weeks, on condition nevertheless that I could then get some other sufficient housekeeper to serve the rest of the time for me—which was, in short, but a very small favour, it being very difficult to get any man to accept of such an employment, that was fit to be entrusted with it.

It is true that shutting up of houses had one effect, which I am sensible was of moment, namely, it confined the distempered people, who would otherwise have been both very troublesome and very dangerous in their running about streets with the distemper upon them—which, when they were delirious, they would have done in a most frightful manner, and as indeed they began to do at first very much, till they were thus restrained; nay, so very open they were that the poor would go about and beg at people's doors, and say they had the plague upon them, and beg rags for their sores, or both, or anything that delirious nature happened to think of.

A poor, unhappy gentlewoman, a substantial citizen's wife, was (if the story be true) murdered by one of these creatures in Aldersgate Street, or that way. He was going along the street, raving mad to be sure, and singing; the people only said he was drunk, but he himself said he had the plague upon him, which it seems was true; and meeting this gentlewoman, he would kiss her. She was terribly frighted, as he was only a rude fellow, and she ran from him, but the street being very thin of people, there was nobody near enough to help her. When she saw he would overtake her, she turned and gave him a thrust so forcibly, he being but weak, and pushed him down backward. But very unhappily, she being so near, he caught hold of her and pulled her down also, and getting up first, mastered her and kissed her; and which was worst of all, when he had done, told her he had the plague, and why should not she have it as well as he? She was frighted enough before, being also young with child; but when she heard him say he had the plague, she screamed out and fell down into a swoon, or in a fit, which, though she recovered a little, yet killed her in a very few days; and I never heard whether she had the plague or no.

Another infected person came and knocked at the door of a citizen's house where they knew him very well; the servant let him in, and being told the master of the house was above, he ran up and came into the room to them as the whole family was at supper. They began to rise up, a little surprised, not knowing what the matter was; but he bid them sit still, he only came to take his leave of them. They asked him, 'Why, Mr—where are you going?' 'Going,' says he; 'I have got the sickness, and shall die tomorrow night.' 'Tis easy to believe, though not to describe, the consternation they were all in. The women and the man's daughters, which were but little girls, were frighted almost to death and got up, one running out at one door and one at another, some downstairs and some upstairs, and getting together as well as they could, locked themselves into their chambers and screamed out at the window for help, as if they had been frighted out of their, wits. The master, more composed than they, though both frighted and provoked, was going to lay hands on him and throw him downstairs, being in a passion; but then, considering a little the condition of the man and the danger of touching him, horror seized his mind, and he stood still like one astonished. The poor distempered man all this while, being as well diseased in his brain as in his body, stood still like one amazed. At length he turns round: 'Ay!' says he, with all the seeming calmness imaginable, 'is it so with you all? Are you all disturbed at me? Why, then I'll e'en go home and die there.' And so he goes immediately downstairs. The servant that had let him in goes down after him with a candle, but was afraid to go past him and open the door, so he stood on the stairs to see what he would do. The man went and opened the door, and went out and flung the door after him. It was some while before the family recovered the fright, but as no ill consequence attended, they have had occasion since to speak of it (You may be sure) with great satisfaction. Though the man was gone, it was some time—nay, as I heard, some days before they recovered themselves of the hurry they were in; nor did they go up and down the house with any assurance till they had burnt a great variety of fumes and perfumes in all the rooms, and made a great many smokes of pitch, of gunpowder, and of sulphur, all separately shifted, and washed their clothes, and the like. As to the poor man, whether he lived or died I don't remember.

It is most certain that, if by the shutting up of houses the sick had not been confined, multitudes who in the height of their fever were delirious and distracted would have been continually running up and down the streets; and even as it was a very great number did so, and offered all sorts of violence to those they met,, even just as a mad dog runs on and bites at every one he meets; nor can I doubt but that, should one of those infected, diseased creatures have bitten any man or woman while the frenzy of the distemper was upon them, they, I mean the person so wounded, would as certainly have been incurably infected as one that was sick before, and had the tokens upon him.

I heard of one infected creature who, running out of his bed in his shirt in the anguish and agony of his swellings, of which he had three upon him, got his shoes on and went to put on his coat; but the nurse resisting, and snatching the coat from him, he threw her down, ran over her, ran downstairs and into the street, directly to the Thames in his shirt; the nurse running after him, and calling to the watch to stop him; but the watchman, frighted at the man, and afraid to touch him, let him go on; upon which he ran down to the Stillyard stairs, threw away his shirt, and plunged into the Thames, and, being a good swimmer, swam quite over the river; and the tide being coming in, as they call it (that is, running westward) he reached the land not till he came about the Falcon stairs, where landing, and finding no people there, it being in the night, he ran about the streets there, naked as he was, for a good while, when, it being by that time high water, he takes the river again, and swam back to the Stillyard, landed, ran up the streets again to his own house, knocking at the door, went up the stairs and into his bed again; and that this terrible experiment cured him of the plague, that is to say, that the violent motion of his arms and legs stretched the parts where the swellings he had upon him were, that is to say, under his arms and his groin, and caused them to ripen and break; and that the cold of the water abated the fever in his blood.

I have only to add that I do not relate this any more than some of the other, as a fact within my own knowledge, so as that I can vouch the truth of them, and especially that of the man being cured by the extravagant adventure, which I confess I do not think very possible; but it may serve to confirm the many desperate things which the distressed people falling into deliriums, and what we call light-headedness, were frequently run upon at that time, and how infinitely more such there would have been if such people had not been confined by the shutting up of houses; and this I take to be the best, if not the only good thing which was performed by that severe method.

On the other hand, the complaints and the murmurings were very bitter against the thing itself. It would pierce the hearts of all that came by to hear the piteous cries of those infected people, who, being thus out of their understandings by the violence of their pain or the heat of their blood, were either shut in or perhaps tied in their beds and chairs, to prevent their doing themselves hurt—and who would make a dreadful outcry at their being confined, and at their being not permitted to die at large, as they called it, and as they would have done before.

This running of distempered people about the streets was very dismal, and the magistrates did their utmost to prevent it; but as it was generally in the night and always sudden when such attempts were made, the officers could not be at band to prevent it; and even when any got out in the day, the officers appointed did not care to meddle with them, because, as they were all grievously infected, to be sure, when they were come to that height, so they were more than ordinarily infectious, and it was one of the most dangerous things that could be to touch them. On the other hand, they generally ran on, not knowing what they did, till they dropped down stark dead, or till they had exhausted their spirits so as that they would fall and then die in perhaps half-an-hour or an hour; and, which was most piteous to hear, they were sure to come to themselves entirely in that half-hour or hour, and then to make most grievous and piercing cries and lamentations in the deep, afflicting sense of the condition they were in. This was much of it before the order for shutting up of houses was strictly put in execution, for at first the watchmen were not so vigorous and severe as they were afterward in the keeping the people in; that is to say, before they were (I mean some of them) severely punished for their neglect, failing in their duty, and letting people who were under their care slip away, or conniving at their going abroad, whether sick or well. But after they saw the officers appointed to examine into their conduct were resolved to have them do their duty or be punished for the omission, they were more exact, and the people were strictly restrained; which was a thing they took so ill and bore so impatiently that their discontents can hardly be described. But there was an absolute necessity for it, that must be confessed, unless some other measures had been timely entered upon, and it was too late for that.

Had not this particular (of the sick being restrained as above) been our case at that time, London would have been the most dreadful place that ever was in the world; there would, for aught I know, have as many people died in the streets as died in their houses; for when the distemper was at its height it generally made them raving and delirious, and when they were so they would never be persuaded to keep in their beds but by force; and many who were not tied threw themselves out of windows when they found they could not get leave to go out of their doors.

It was for want of people conversing one with another, in this time of calamity, that it was impossible any particular person could come at the knowledge of all the extraordinary cases that occurred in different families; and particularly I believe it was never known to this day how many people in their deliriums drowned themselves in the Thames, and in the river which runs from the marshes by Hackney, which we generally called Ware River, or Hackney River. As to those which were set down in the weekly bill, they were indeed few; nor could it be known of any of those whether they drowned themselves by accident or not. But I believe I might reckon up more who within the compass of my knowledge or observation really drowned themselves in that year, than are put down in the bill of all put together: for many of the bodies were never found who yet were known to be lost; and the like in other methods of self-destruction. There was also one man in or about Whitecross Street burned himself to death in his bed; some said it was done by himself, others that it was by the treachery of the nurse that attended him; but that he had the plague upon him was agreed by all.

It was a merciful disposition of Providence also, and which I have many times thought of at that time, that no fires, or no considerable ones at least, happened in the city during that year, which, if it had been otherwise, would have been very dreadful; and either the people must have let them alone unquenched, or have come together in great crowds and throngs, unconcerned at the danger of the infection, not concerned at the houses they went into, at the goods they handled, or at the persons or the people they came among. But so it was, that excepting that in Cripplegate parish, and two or three little eruptions of fires, which were presently extinguished, there was no disaster of that kind happened in the whole year. They told us a story of a house in a place called Swan Alley, passing from Goswell Street, near the end of Old Street, into St John Street, that a family was infected there in so terrible a manner that every one of the house died. The last person lay dead on the floor, and, as it is supposed, had lain herself all along to die just before the fire; the fire, it seems, had fallen from its place, being of wood, and had taken hold of the boards and the joists they lay on, and burnt as far as just to the body, but had not taken hold of the dead body (though she had little more than her shift on) and had gone out of itself, not burning the rest of the house, though it was a slight timber house. How true this might be I do not determine, but the city being to suffer severely the next year by fire, this year it felt very little of that calamity.

Indeed, considering the deliriums which the agony threw people into, and how I have mentioned in their madness, when they were alone, they did many desperate things, it was very strange there were no more disasters of that kind.

It has been frequently asked me, and I cannot say that I ever knew how to give a direct answer to it, how it came to pass that so many infected people appeared abroad in the streets at the same time that the houses which were infected were so vigilantly searched, and all of them shut up and guarded as they were.

I confess I know not what answer to give to this, unless it be this: that in so great and populous a city as this is it was impossible to discover every house that was infected as soon as it was so, or to shut up all the houses that were infected; so that people had the liberty of going about the streets, even where they Pleased, unless they were known to belong to such-and- such infected houses.

It is true that, as several physicians told my Lord Mayor, the fury of the contagion was such at some particular times, and people sickened so fast and died so soon, that it was impossible, and indeed to no purpose, to go about to inquire who was sick and who was well, or to shut them up with such exactness as the thing required, almost every house in a whole street being infected, and in many places every person in some of the houses; and that which was still worse, by the time that the houses were known to be infected, most of the persons infected would be stone dead, and the rest run away for fear of being shut up; so that it was to very small purpose to call them infected houses and shut them up, the infection having ravaged and taken its leave of the house before it was really known that the family was any way touched.

This might be sufficient to convince any reasonable person that as it was not in the power of the magistrates or of any human methods of policy, to prevent the spreading the infection, so that this way of shutting up of houses was perfectly insufficient for that end. Indeed it seemed to have no manner of public good in it, equal or proportionable to the grievous burden that it was to the particular families that were so shut up; and, as far as I was employed by the public in directing that severity, I frequently found occasion to see that it was incapable of answering the end. For example, as I was desired, as a visitor or examiner, to inquire into the particulars of several families which were infected, we scarce came to any house where the plague had visibly appeared in the family but that some of the family were fled and gone. The magistrates would resent this, and charge the examiners with being remiss in their examination or inspection. But by that means houses were long infected before it was known. Now, as I was in this dangerous office but half the appointed time, which was two months, it was long enough to inform myself that we were no way capable of coming at the knowledge of the true state of any family but by inquiring at the door or of the neighbours. As for going into every house to search, that was a part no authority would offer to impose on the inhabitants, or any citizen would undertake: for it would have been exposing us to certain infection and death, and to the ruin of our own families as well as of ourselves; nor would any citizen of probity, and that could be depended upon, have stayed in the town if they had been made liable to such a severity.

Seeing then that we could come at the certainty of things by no method but that of inquiry of the neighbours or of the family, and on that we could not justly depend, it was not possible but that the uncertainty of this matter would remain as above.

It is true masters of families were bound by the order to give notice to the examiner of the place wherein he lived, within two hours after he should discover it, of any person being sick in his house (that is to say, having signs of the infection) but they found so many ways to evade this and excuse their negligence that they seldom gave that notice till they had taken measures to have every one escape out of the house who had a mind to escape, whether they were sick or sound; and while this was so, it is easy to see that the shutting up of houses was no way to be depended upon as a sufficient method for putting a stop to the infection because, as I have said elsewhere, many of those that so went out of those infected houses had the plague really upon them, though they might really think themselves sound. And some of these were the people that walked the streets till they fell down dead, not that they were suddenly struck with the distemper as with a bullet that killed with the stroke, but that they really had the infection in their blood long before; only, that as it preyed secretly on the vitals, it appeared not till it seized the heart with a mortal power, and the patient died in a moment, as with a sudden fainting or an apoplectic fit.

I know that some even of our physicians thought for a time that those people that so died in the streets were seized but that moment they fell, as if they had been touched by a stroke from heaven as men are killed by a flash of lightning—but they found reason to alter their opinion afterward; for upon examining the bodies of such after they were dead, they always either had tokens upon them or other evident proofs of the distemper having been longer upon them than they had otherwise expected.

This often was the reason that, as I have said, we that were examiners were not able to come at the knowledge of the infection being entered into a house till it was too late to shut it up, and sometimes not till the people that were left were all dead. In Petticoat Lane two houses together were infected, and several people sick; but the distemper was so well concealed, the examiner, who was my neighbour, got no knowledge of it till notice was sent him that the people were all dead, and that the carts should call there to fetch them away. The two heads of the families concerted their measures, and so ordered their matters as that when the examiner was in the neighbourhood they appeared generally at a time, and answered, that is, lied, for one another, or got some of the neighbourhood to say they were all in health—and perhaps knew no better—till, death making it impossible to keep it any longer as a secret, the dead-carts were called in the night to both the houses t and so it became public. But when the examiner ordered the constable to shut up the houses there was nobody left in them but three people, two in one house and one in the other, just dying, and a nurse in each house who acknowledged that they had buried five before, that the houses had been infected nine or ten days, and that for all the rest of the two families, which were many, they were gone, some sick, some well, or whether sick or well could not be known.

In like manner, at another house in the same lane, a man having his family infected but very unwilling to be shut up, when he could conceal it no longer, shut up himself; that is to say, he set the great red cross upon his door with the words, 'Lord have mercy upon us', and so deluded the examiner, who supposed it had been done by the constable by order of the other examiner, for there were two examiners to every district or precinct. By this means he had free egress and regress into his house again, and out of it, as he pleased, notwithstanding it was infected, till at length his stratagem was found out; and then he, with the sound part of his servants and family, made off and escaped, so they were not shut up at all.

These things made it very hard, if not impossible, as I have said, to prevent the spreading of an infection by the shutting up of houses unless the people would think the shutting of their houses no grievance, and be so willing to have it done as that they would give notice duly and faithfully to the magistrates of their being infected as soon as it was known by themselves; but as that cannot be expected from them, and the examiners cannot be supposed, as above, to go into their houses to visit and search, all the good of shutting up houses will be defeated, and few houses will be shut up in who will be discovered by the terror and consternation which the things put them into.

I got myself discharged of the dangerous office I was in as soon as I could get another admitted, whom I had obtained for a little money to accept of it; and so, instead of serving the two months, which was directed, I was not above three weeks in it; and a great while too, considering it was in the month of August, at which time the distemper began to rage with great violence at our end of the town.

In the execution of this office I could not refrain speaking my opinion among my neighbours as to this shutting up the people in their houses; in which we saw most evidently the severities that were used, though grievous in themselves, had also this particular objection against them: namely, that they did not answer the end, as I have said, but that the distempered people went day by day about the streets; and it was our united opinion that a method to have removed the sound from the sick, in case of a particular house being visited, would have been much more reasonable on many accounts, leaving nobody with the sick persons but such as should on such occasion request to stay and declare themselves content to be shut up with them

Our scheme for removing those that were sound from those that were sick was only in such houses as were infected, and confining the sick was no confinement; those that could not stir would not complain while they were in their senses and while they had the power of judging. Indeed, when they came to be delirious and light-headed, then they would cry out of the cruelty of being confined; but for the removal of those that were well, we thought it highly reasonable and just, for their own sakes, they should be removed from the sick, and that for other people's safety they should keep retired for a while, to see that they were sound, and might not infect others; and we thought twenty or thirty days enough for this.

Now, certainly, if houses had been provided on purpose for those that were sound to perform this demi-quarantine in, they would have much less reason to think themselves injured in such a restraint than in being confined with infected people in the houses where they lived.

It is here, however, to be observed that after the funerals became so many that people could not toll the bell, mourn or weep, or wear black for one another, as they did before; no, nor so much as make coffins for those that died; so after a while the fury of the infection appeared to be so increased that, in short, they shut up no houses at all. It seemed enough that all the remedies of that

Kind had been used till they were found fruitless, and that the plague spread itself with an irresistible fury; so that as the fire the succeeding year spread itself, and burned with such violence that the citizens, in despair, gave over their endeavours to extinguish it, so in the plague it came at last to such violence that the people sat still looking at one another, and seemed quite abandoned to despair; whole streets seemed to be desolated, and not to be shut up only, but to be emptied of their inhabitants; doors were left open, windows stood shattering with the wind in empty houses for want of people to shut them. In a word, people began to give up themselves to their fears and to think that all regulations and methods were in vain, and that there was nothing to be hoped for but an universal desolation; and it was even in the height of this general despair that it Pleased God to stay His hand, and to slacken the fury of the contagion in such a manner as was even surprising, like its beginning, and demonstrated it to be His own particular hand, and that above, if not without the agency of means, as I shall take notice of in its proper place.

But I must still speak of the plague as in its height, raging even to desolation, and the people under the most dreadful consternation, even, as I have said, to despair. It is hardly credible to what excess the passions of men carried them in this extremity of the distemper, and this part, I think, was as moving as the rest. What could affect a man in his full power of reflection, and what could make deeper impressions on the soul, than to see a man almost naked, and got out of his house, or perhaps out of his bed, into the street, come out of Harrow Alley, a populous conjunction or collection of alleys, courts, and passages in the Butcher Row in Whitechappel—I say, what could be more affecting than to see this poor man come out into the open street, run dancing and singing and making a thousand antic gestures, with five or six women and children running after him, crying and calling upon him for the Lord's sake to come back, and entreating the help of others to bring him back, but all in vain, nobody daring to lay a hand upon him or to come near him?

This was a most grievous and afflicting thing to me, who saw it all from my own windows; for all this while the poor afflicted man was, as I observed it, even then in the utmost agony of pain, having (as they said) two swellings upon him which could not be brought to break or to suppurate; but, by laying strong caustics on them, the surgeons had, it seems, hopes to break them—which caustics were then upon him, burning his flesh as with a hot iron. I cannot say what became of this poor man, but I think he continued roving about in that manner till he fell down and died.

No wonder the aspect of the city itself was frightful. The usual concourse of people in the streets, and which used to be supplied from our end of the town, was abated. The Exchange was not kept shut, indeed, but it was no more frequented. The fires were lost; they had been almost extinguished for some days by a very smart and hasty rain. But that was not all; some of the physicians insisted that they were not only no benefit, but injurious to the health of people. This they made a loud clamour about, and complained to the Lord Mayor about it. On the other hand, others of the same faculty, and eminent too, opposed them, and gave their reasons why the fires were, and must be, useful to assuage the violence of the distemper. I cannot give a full account of their arguments on both sides; only this I remember, that they cavilled very much with one another. Some were for fires, but that they must be made of wood and not coal, and of particular sorts of wood too, such as fir in particular, or cedar, because of the strong effluvia of turpentine; others were for coal and not wood, because of the sulphur and bitumen; and others were for neither one or other. Upon the whole, the Lord Mayor ordered no more fires, and especially on this account, namely, that the plague was so fierce that they saw evidently it defied all means, and rather seemed to increase than decrease upon any application to check and abate it; and yet this amazement of the magistrates proceeded rather from want of being able to apply any means successfully than from any unwillingness either to expose themselves or undertake the care and weight of business; for, to do them justice, they neither spared their pains nor their persons. But nothing answered; the infection raged, and the people were now frighted and terrified to the last degree: so that, as I may say, they gave themselves up, and, as I mentioned above, abandoned themselves to their despair.

But let me observe here that, when I say the people abandoned themselves to despair, I do not mean to what men call a religious despair, or a despair of their eternal state, but I mean a despair of their being able to escape the infection or to outlive the plague, which they saw was so raging and so irresistible in its force that indeed few people that were touched with it in its height, about August and September, escaped; and, which is very particular, contrary to its ordinary operation in June and July, and the beginning of August, when, as I have observed, many were infected, and continued so many days, and then went off after having had the poison in their blood a long time; but now, on the contrary, most of the people who were taken during the two last weeks in August and in the three first weeks in September, generally died in two or three days at furthest, and many the very same day they were taken; whether the dog-days, or, as our astrologers pretended to express themselves, the influence of the dog-star, had that malignant effect, or all those who had the seeds of infection before in them brought it up to a maturity at that time altogether, I know not; but this was the time when it was reported that above 3000 people died in one night; and they that would have us believe they more critically observed it pretend to say that they all died within the space of two hours, viz., between the hours of one and three in the morning.

As to the suddenness of people's dying at this time, more than before, there were innumerable instances of it, and I could name several in my neighbourhood. One family without the Bars, and not far from me, were all seemingly well on the Monday, being ten in family. That evening one maid and one apprentice were taken ill and died the next morning—when the other apprentice and two children were touched, whereof one died the same evening, and the other two on Wednesday. In a word, by Saturday at noon the master, mistress, four children, and four servants were all gone, and the house left entirely empty, except an ancient woman who came in to take charge of the goods for the master of the family's brother, who lived not far off, and who had not been sick.

Many houses were then left desolate, all the people being carried away dead, and especially in an alley farther on the same side beyond the Bars, going in at the sign of Moses and Aaron, there were several houses together which, they said, had not one person left alive in them; and some that died last in several of those houses were left a little too long before they were fetched out to be buried; the reason of which was not, as some have written very untruly, that the living were not sufficient to bury the dead, but that the mortality was so great in the yard or alley that there was nobody left to give notice to the buriers or sextons that there were any dead bodies there to be buried. It was said, how true I know not, that some of those bodies were so much corrupted and so rotten that it was with difficulty they were carried; and as the carts could not come any nearer than to the Alley Gate in the High Street, it was so much the more difficult to bring them along; but I am not certain how many bodies were then left. I am sure that ordinarily it was not so.

As I have mentioned how the people were brought into a condition to despair of life and abandon themselves, so this very thing had a strange effect among us for three or four weeks; that is, it made them bold and venturous: they were no more shy of one another, or restrained within doors, but went anywhere and everywhere, and began to converse. One would say to another, 'I do not ask you how you are, or say how I am; it is certain we shall all go; so 'tis no matter who is all sick or who is sound'; and so they ran desperately into any place or any company.

As it brought the people into public company, so it was surprising how it brought them to crowd into the churches. They inquired no more into whom they sat near to or far from, what offensive smells they met with, or what condition the people seemed to be in; but, looking upon themselves all as so many dead corpses, they came to the churches without the least caution, and crowded together as if their lives were of no consequence compared to the work which they came about there. Indeed, the zeal which they showed in coming, and the earnestness and affection they showed in their attention to what they heard, made it manifest what a value people would all put upon the worship of God if they thought every day they attended at the church that it would be their last.

Nor was it without other strange effects, for it took away, all manner of prejudice at or scruple about the person whom they found in the pulpit when they came to the churches. It cannot be doubted but that many of the ministers of the parish churches were cut off, among others, in so common and dreadful a calamity; and others had not courage enough to stand it, but removed into the country as they found means for escape. As then some parish churches were quite vacant and forsaken, the people made no scruple of desiring such Dissenters as had been a few years before deprived of their livings by virtue of the Act of Parliament called the Act of Uniformity to preach in the churches; nor did the church ministers in that case make any difficulty of accepting their assistance; so that many of those whom they called silenced ministers had their mouths opened on this occasion and preached publicly to the people.

Here we may observe and I hope it will not be amiss to take notice of it that a near view of death would soon reconcile men of good principles one to another, and that it is chiefly owing to our easy situation in life and our putting these things far from us that our breaches are fomented, ill blood continued, prejudices, breach of charity and of Christian union, so much kept and so far carried on among us as it is. Another plague year would reconcile all these differences; a dose conversing with death, or with diseases that threaten death, would scum off the gall from our tempers, remove the animosities among us, and bring us to see with differing eyes than those which we looked on things with before. As the people who had been used to join with the Church were reconciled at this time with the admitting the Dissenters to preach to them, so the Dissenters, who with an uncommon prejudice had broken off from the communion of the Church of England, were now content to come to their parish churches and to conform to the worship which they did not approve of before; but as the terror of the infection abated, those things all returned again to their less desirable channel and to the course they were in before.

I mention this but historically. I have no mind to enter into arguments to move either or both sides to a more charitable compliance one with another. I do not see that it is probable such a discourse would be either suitable or successful; the breaches seem rather to widen, and tend to a widening further, than to closing, and who am I that I should think myself able to influence either one side or other? But this I may repeat again, that 'tis evident death will reconcile us all; on the other side the grave we shall be all brethren again. In heaven, whither I hope we may come from all parties and persuasions, we shall find neither prejudice or scruple; there we shall be of one principle and of one opinion. Why we cannot be content to go hand in hand to the Place where we shall join heart and hand without the least hesitation, and with the most complete harmony and affection—I say, why we cannot do so here I can say nothing to, neither shall I say anything more of it but that it remains to be lamented.

I could dwell a great while upon the calamities of this dreadful time, and go on to describe the objects that appeared among us every day, the dreadful extravagancies which the distraction of sick people drove them into; how the streets began now to be fuller of frightful objects, and families to be made even a terror to themselves. But after I have told you, as I have above, that one man, being tied in his bed, and finding no other way to deliver himself, set the bed on fire with his candle, which unhappily stood within his reach, and burnt himself in his bed; and how another, by the insufferable torment he bore, danced and sung naked in the streets, not knowing one ecstasy from another; I say, after I have mentioned these things, what can be added more? What can be said to represent the misery of these times more lively to the reader, or to give him a more perfect idea of a complicated distress?

I must acknowledge that this time was terrible, that I was sometimes at the end of all my resolutions, and that I had not the courage that I had at the beginning. As the extremity brought other people abroad, it drove me home, and except having made my voyage down to Blackwall and Greenwich, as I have related, which was an excursion, I kept afterwards very much within doors, as I had for about a fortnight before. I have said already that I repented several times that I had ventured to stay in town, and had not gone away with my brother and his family, but it was too late for that now; and after I had retreated and stayed within doors a good while before my impatience led me abroad, then they called me, as I have said, to an ugly and dangerous office which brought me out again; but as that was expired while the height of the distemper lasted, I retired again, and continued dose ten or twelve days more, during which many dismal spectacles represented themselves in my view out of my own windows and in our own street—as that particularly from Harrow Alley, of the poor outrageous creature which danced and sung in his agony; and many others there were. Scarce a day or night passed over but some dismal thing or other happened at the end of that Harrow Alley, which was a place full of poor people, most of them belonging to the butchers or to employments depending upon the butchery.

Sometimes heaps and throngs of people would burst out of the alley, most of them women, making a dreadful clamour, mixed or compounded of screeches, cryings, and calling one another, that we could not conceive what to make of it. Almost all the dead part of the night the dead-cart stood at the end of that alley, for if it went in it could not well turn again, and could go in but a little way. There, I say, it stood to receive dead bodies, and as the churchyard was but a little way off, if it went away full it would soon be back again. It is impossible to describe the most horrible cries and noise the poor people would make at their bringing the dead bodies of their children and friends out of the cart, and by the number one would have thought there had been none left behind, or that there were people enough for a small city living in those places. Several times they cried 'Murder', sometimes 'Fire'; but it was easy to perceive it was all distraction, and the complaints of distressed and distempered people.

I believe it was everywhere thus as that time, for the plague raged for six or seven weeks beyond all that I have expressed, and came even to such a height that, in the extremity, they began to break into that excellent order of which I have spoken so much in behalf of the magistrates; namely, that no dead bodies were seen in the street or burials in the daytime: for there was a necessity in this extremity to bear with its being otherwise for a little while.

One thing I cannot omit here, and indeed I thought it was extraordinary, at least it seemed a remarkable hand of Divine justice: viz., that all the predictors, astrologers, fortune-tellers, and what they called cunning-men, conjurers, and the like: calculators of nativities and dreamers of dream, and such people, were gone and vanished; not one of them was to be found. I am verily persuaded that a great number of them fell in the heat of the calamity, having ventured to stay upon the prospect of getting great estates; and indeed their gain was but too great for a time, through the madness and folly of the people. But now they were silent; many of them went to their long home, not able to foretell their own fate or to calculate their own nativities. Some have been critical enough to say that every one of them died. I dare not affirm that; but this I must own, that I never heard of one of them that ever appeared after the calamity was over.

But to return to my particular observations during this dreadful part of the visitation. I am now come, as I have said, to the month of September, which was the most dreadful of its kind, I believe, that ever London saw; for, by all the accounts which I have seen of the preceding visitations which have been in London, nothing has been like it, the number in the weekly bill amounting to almost 40,000 from the 22nd of August to the 26th of September, being but five weeks. The particulars of the bills are as follows, viz.:—

From August the 22nd to the 29th 7496
" " 29th " 5th September 8252
" 12th " 19th 8297
" " 19th " 26th 6460

38,195

This was a prodigious number of itself, but if I should add the reasons which I have to believe that this account was deficient, and how deficient it was, you would, with me, make no scruple to beheve that there died above ten thousand a week for all those weeks, one week with another, and a proportion for several weeks both before and after. The confusion among the people, especially within the city, at that time, was inexpressible. The terror was so great at last that the courage of the people appointed to carry away the dead began to fail them; nay, several of them died, although they had the distemper before and were recovered, and some of them dropped down when they have been carrying the bodies even at the pit side, and just ready to throw them in; and this confusion was greater in the city because they had flattered themselves with hopes of escaping, and thought the bitterness of death was past. One cart, they told us, going up Shoreditch was forsaken of the drivers, or being left to one man to drive, he died in the street; and the horses going on overthrew the cart, and left the bodies, some thrown out here, some there, in a dismal manner. Another cart was, it seems, found in the great pit in Finsbury Fields, the driver being dead, or having been gone and abandoned it, and the horses running too near it, the cart fell in and drew the horses in also. It was suggested that the driver was thrown in with it and that the cart fell upon him, by reason his whip was seen to be in the pit among the bodies; but that, I suppose, could not be certain.

In our parish of Aldgate the dead-carts were several times, as I have heard, found standing at the churchyard gate full of dead bodies, but neither bellman or driver or any one else with it; neither in these or many other cases did they know what bodies they had in their cart, for sometimes they were let down with ropes out of balconies and out of windows, and sometimes the bearers brought them to the cart, sometimes other people; nor, as the men themselves said, did they trouble themselves to keep any account of the numbers.

The vigilance of the magistrates was now put to the utmost trial—and, it must be confessed, can never be enough acknowledged on this occasion also; whatever expense or trouble they were at, two things were never neglected in the city or suburbs either:—

(1) Provisions were always to be had in full plenty, and the price not much raised neither, hardly worth speaking. ,

(2) No dead bodies lay unburied or uncovered; and if one walked from one end of the city to another, no funeral or sign of it was to be seen in the daytime, except a little, as I have said above, in the three first weeks in September.

This last article perhaps will hardly be believed when some accounts which others have published since that shall be seen, wherein they say that the dead lay unburied, which I am assured was utterly false; at least, if it had been anywhere so, it must have been in houses where the living were gone from the dead (having found means, as I have observed, to escape) and where no notice was given to the officers. All which amounts to nothing at all in the case in hand; for this I am positive in, having myself been employed a little in the direction of that part in the parish in which I lived, and where as great a desolation was made in proportion to the number of inhabitants as was anywhere; I say, I am sure that there were no dead bodies remained unburied; that is to say, none that the proper officers knew of; none for want of people to carry them off, and buriers to put them into the ground and cover them; and this is sufficient to the argument; for what might lie in houses and holes, as in Moses and Aaron

Alley, is nothing; for it is most certain they were buried as soon as they were found. As to the first article (namely, of provisions, the scarcity or dearness), though I have mentioned it before and shall speak of it again, yet I must observe here:—

(1) The price of bread in particular was not much raised; for in the beginning of the year, viz., in the first week in March, the penny wheaten loaf was ten ounces and a half; and in the height of the contagion it was to be had at nine ounces and a half, and never dearer, no, not all that season. And about the beginning of November it was sold ten ounces and a half again; the like of which, I believe, was never heard of in any city, under so dreadful a visitation, before.

(2) Neither was there (which I wondered much at) any want of bakers or ovens kept open to supply the people with the bread; but this was indeed alleged by some families, viz., that their maidservants, going to the bakehouses with their dough to be baked, which was then the custom, sometimes came home with the sickness (that is to say the plague) upon them.

In all this dreadful visitation there were, as I have said before, but two pest- houses made use of, viz., one in the fields beyond Old Street and one in Westminster; neither was there any compulsion used in carrying people thither. Indeed there was no need of compulsion in the case, for there were thousands of poor distressed people who, having no help or conveniences or supplies but of charity, would have been very glad to have been carried thither and been taken care of; which, indeed, was the only thing that I think was wanting in the whole public management of the city, seeing nobody was here allowed to be brought to the pest-house but where money was given, or security for money, either at their introducing or upon their being cured and sent out—for very many were sent out again whole; and very good physicians were appointed to those places, so that many people did very well there, of which I shall make mention again. The principal sort of people sent thither were, as I have said, servants who got the distemper by going of errands to fetch necessaries to the families where they lived, and who in that case, if they came home sick, were removed to preserve the rest of the house; and they were so well looked after there in all the time of the visitation that there was but 156 buried in all at the London pest-house, and 159 at that of Westminster.

By having more pest-houses I am far from meaning a forcing all people into such places. Had the shutting up of houses been omitted and the sick hurried out of their dwellings to pest-houses, as some proposed, it seems, at that time as well as since, it would certainly have been much worse than it was. The very removing the sick would have been a spreading of the infection, and the rather because that removing could not effectually clear the house where the sick person was of the distemper; and the rest of the family, being then left at liberty, would certainly spread it among others.

The methods also in private families, which would have been universally used to have concealed the distemper and to have concealed the persons being sick, would have been such that the distemper would sometimes have seized a whole family before any visitors or examiners could have known of it. On the other hand, the prodigious numbers which would have been sick at a time would have exceeded all the capacity of public pest-houses to receive them, or of public officers to discover and remove them.

This was well considered in those days, and I have heard them talk of it often. The magistrates had enough to do to bring people to submit to having their houses shut up, and many ways they deceived the watchmen and got out, as I have observed. But that difficulty made it apparent that they would have found it impracticable to have gone the other way to work, for they could never have forced the sick people out of their beds and out of their dwellings. It must not have been my Lord Mayor's officers, but an army of officers, that must have attempted it; and tile people, on the other hand, would have been enraged and desperate, and would have killed those that should have offered to have meddled with them or with their children and relations, whatever had befallen them for it; so that they would have made the people, who, as it was, were in the most terrible distraction imaginable,

I say, they would have made them stark mad; whereas the magistrates found it proper on several accounts to treat them with lenity and compassion, and not with violence and terror, such as dragging the sick out of their houses or obliging them to remove themselves, would have been.

This leads me again to mention the time when the plague first began; that is to say, when it became certain that it would spread over the whole town, when, as I have said, the better sort of people first took the alarm and began to hurry themselves out of town. It was true, as I observed in its place, that the throng was so great, and the coaches, horses, waggons, and carts were so many, driving and dragging the people away, that it looked as if all the city was running away; and had any regulations been published that had been terrifying at that time, especially such as would pretend to dispose of the people otherwise than they would dispose of themselves, it would have put both the city and suburbs into the utmost confusion.

But the magistrates wisely caused the people to be encouraged, made very good bye-laws for the regulating the citizens, keeping good order in the streets, and making everything as eligible as possible to all sorts of people.

In the first place, the Lord Mayor and the sheriffs, the Court of Aldermen, and a certain number of the Common Council men, or their deputies, came to a resolution and published it, viz., that they would not quit the city themselves, but that they would be always at hand for the preserving good order in every place and for the doing justice on all occasions; as also for the distributing the public charity to the poor; and, in a word, for the doing the duty and discharging the trust reposed in them by the citizens to the utmost of their power.

In pursuance of these orders, the Lord Mayor, sheriffs, &c., held councils every day, more or less, for making such dispositions as they found needful for preserving the civil peace; and though they used the people with all possible gentleness and clemency, yet all manner of presumptuous rogues such as thieves, housebreakers, plunderers of the dead or of the sick, were duly punished, and several declarations were continually published by the Lord Mayor and Court of Aldermen against such.

Also all constables and churchwardens were enjoined to stay in the city upon severe penalties, or to depute such able and sufficient housekeepers as the deputy aldermen or Common Council men of the precinct should approve, and for whom they should give security; and also security in case of mortality that they would forthwith constitute other constables in their stead.

These things re-established the minds of the people very much, especially in the first of their fright, when they talked of making so universal a flight that the city would have been in danger of being entirely deserted of its inhabitants except the poor, and the country of being plundered and laid waste by the multitude. Nor were the magistrates deficient in performing their part as boldly as they promised it; for my Lord Mayor and the sheriffs were continually in the streets and at places of the greatest danger, and though they did not care for having too great a resort of people crowding about them, yet in emergent cases they never denied the people access to them, and heard with patience all their grievances and complaints. My Lord Mayor had a low gallery built on purpose in his hall, where he stood a little removed from the crowd when any complaint came to be heard, that he might appear with as much safety as possible.

Likewise the proper officers, called my Lord Mayor's officers, constantly attended in their turns, as they were in waiting; and if any of them were sick or infected, as some of them were, others were instantly employed to fill up and officiate in their places till it was known whether the other should live or die.

In like manner the sheriffs and aldermen did in their several stations and wards, where they were placed by office, and the sheriff's officers or sergeants were appointed to receive orders from the respective aldermen in their turn, so that justice was executed in all cases without interruption. In the next place, it was one of their particular cares to see the orders for the freedom of the markets observed, and in this part either the Lord Mayor or one or both of the sheriffs were every market-day on horseback to see their orders executed and to see that the country people had all possible encouragement and freedom in their coming to the markets and going back again, and that no nuisances or frightful objects should be seen in the streets to terrify them or make them unwilling to come. Also the bakers were taken under particular order, and the Master of the Bakers' Company was, with his court of assistants, directed to see the order of my Lord Mayor for their regulation put in execution, and the due assize of bread (which was weekly appointed by my Lord Mayor) observed; and all the bakers were obliged to keep their oven going constantly, on pain of losing the privileges of a freeman of the city of London.

By this means bread was always to be had in plenty, and as cheap as usual, as I said above; and provisions were never wanting in the markets, even to such a degree that I often wondered at it, and reproached myself with being so timorous and cautious in stirring abroad, when the country people came freely and boldly to market, as if there had been no manner of infection in the city, or danger of catching it.

It was indeed one admirable piece of conduct in the said magistrates that the streets were kept constantly clear and free from all manner of frightful objects, dead bodies, or any such things as were indecent or unpleasant—unless where anybody fell down suddenly or died in the streets, as I have said above; and these were generally covered with some cloth or blanket, or removed into the next churchyard till night. All the needful works that carried terror with them, that were both dismal and dangerous, were done in the night; if any diseased bodies were removed, or dead bodies buried, or infected clothes burnt, it was done in the night; and all the bodies which were thrown into the great pits in the several churchyards or burying- grounds, as has been observed, were so removed in the night, and everything was covered and closed before day. So that in the daytime there was not the least signal of the calamity to be seen or heard of, except what was to be observed from the emptiness of the streets, and sometimes from the passionate outcries and lamentations of the people, out at their windows, and from the numbers of houses and shops shut up.

Nor was the silence and emptiness of the streets so much in the city as in the out-parts, except just at one particular time when, as I have mentioned, the plague came east and spread over all the city. It was indeed a merciful disposition of God, that as the plague began at one end of the town first (as has been observed at large) so it proceeded progressively to other parts, and did not come on this way, or eastward, till it had spent its fury in the West part of the town; and so, as it came on one way, it abated another. For example, it began at St Giles's and the Westminster end of the town, and it was in its height in all that part by about the middle of July, viz., in St Giles- in- the-Fields, St Andrew's, Holborn, St Clement Danes, St

Martin-in-the-Fields, and in Westminster. The latter end of July it decreased in those parishes; and coming east, it increased prodigiously in Cripplegate, St Sepulcher's, St James's, Clarkenwell, and St Bride's and Aldersgate. While it was in all these parishes, the city and all the parishes of the Southwark side of the water and all Stepney, Whitechappel, Aldgate, Wapping, and Ratcliff, were very little touched; so that people went about their business unconcerned, carried on their trades, kept open their shops, and conversed freely with one another in all the city, the east and north-east suburbs, and in Southwark, almost as if the plague had not been among us.

Even when the north and north-west suburbs were fully infected, viz., Cripplegate, Clarkenwell, Bishopsgate, and Shoreditch, yet still all the rest were tolerably well. For example from 25th July to 1st August the bill stood thus of all diseases:—

St Giles, Cripplegate 554
St Sepulchers 250
Clarkenwell 103
Bishopsgate 116
Shoreditch 110
Stepney parish 127
Aldgate 92
Whitechappel 104
All the ninety-seven parishes within the walls 228
All the parishes in Southwark 205
Total 1889

So that, in short, there died more that week in the two parishes of Cripplegate and St Sepulcher by forty-eight than in all the city, all the east suburbs, and all the Southwark parishes put together. This caused the reputation of the city's health to continue all over England—and especially in the counties and markets adjacent, from whence our supply of provisions chiefly came even much longer than that health itself continued; for when the people came into the streets from the country by Shoreditch and Bishopsgate, or by Old Street and Smithfield, they would see the out-streets empty and the houses and shops shut, and the few people that were stirring there walk in the middle of the streets. But when they came within the city, there things looked better, and the markets and shops were open, and the people walking about the streets as usual, though not quite so many; and this continued till the latter end of August and the beginning of September.

But then the case altered quite; the distemper abated in the west and northwest parishes, and the weight of the infection lay on the city and the eastern suburbs, and the Southwark side, and this in a frightful manner.

Then, indeed, the city began to look dismal, shops to be shut, and the streets desolate. In the High Street, indeed, necessity made people stir abroad on many occasions; and there would be in the middle of the day a pretty many people, but in the mornings and evenings scarce any to be seen, even there, no, not in Cornhill and Cheapside.

These observations of mine were abundantly confirmed by the weekly bills of mortality for those weeks, an abstract of which, as they respect the parishes which. I have mentioned and as they make the calculations I speak of very evident, take as follows.

The weekly bill, which makes out this decrease of the burials in the west and north side of the city, stands thus:—

From the 12th of September to the 19th:—
St Giles, Cripplegate 456
St Giles-in-the-Fields 140
Clarkenwell 77

St Sepulcher 214
St Leonard, Shoreditch 183
Stepney parish 716
Aldgate 623
Whitechappel 532
In the ninety-seven parishes within the walls 1493
In the eight parishes on Southwark side 1636
Total 6060

Here is a strange change of things indeed, and a sad change it was; and had it held for two months more than it did, very few people would have been left alive. But then such, I say, was the merciful disposition of God that, when it was thus, the west and north part which had been so dreadfully visited at first, grew, as you see, much better; and as the people disappeared here, they began to look abroad again there; and the next week or two altered it still more; that is, more to the encouragement of tile other part of the town. For example:—

From the 19th of September to the 26th:—
St Giles, Cripplegate 277
St Giles-in-the-Fields 119
Clarkenwell 76
St Sepulchers 193
St Leonard, Shoreditch 146
Stepney parish 616
Aldgate 496
Whitechappel 346
In the ninety-seven parishes within the walls 1268
In the eight parishes on Southwark side 1390
Total 4927
From the 26th of September to the 3rd of October:-
St Giles, Cripplegate 196

St Giles-in-the-Fields 95
Clarkenwell 48
St Sepulchers 137
St Leonard, Shoreditch 128
Stepney parish 674
Aldgate 372
Whitechappel 328
In the ninety-seven parishes within the walls 1149
In the eight parishes on Southwark side 1201
Total 4382

And now the misery of the city and of the said east and south parts was complete indeed; for, as you see, the weight of the distemper lay upon those parts, that is to say, the city, the eight parishes over the river, with the parishes of Aldgate, Whitechappel, and Stepney; and this was the time that the bills came up to such a monstrous height as that I mentioned before, and that eight or nine, and, as I believe, ten or twelve thousand a week, died; for it is my settled opinion that they never could come at any just account of the numbers, for the reasons which I have given already.

Nay, one of the most eminent physicians, who has since published in Latin an account of those times, and of his observations says that in one week there died twelve thousand people, and that particularly there died four thousand in one night; though I do not remember that there ever was any such particular night so remarkably fatal as that such a number died in it. However, all this confirms what I have said above of the uncertainty of the bills of mortality, &c., of which I shall say more hereafter.

And here let me take leave to enter again, though it may seem a repetition of circumstances, into a description of the miserable condition of the city itself, and of those parts where I lived at this particular time. The city and those other parts, notwithstanding the great numbers of people that were gone into the country, was vastly full of people; and perhaps the fuller because people had for a long time a strong belief that the plague would not come into the city, nor into Southwark, no, nor into Wapping or Ratcliff at all; nay, such was the assurance of the people on that head that many removed from the suburbs on the west and north sides, into those eastern and south sides as for safety; and, as I verily believe, carried the plague amongst them there perhaps sooner than they would otherwise have had it.

Here also I ought to leave a further remark for the use of posterity, concerning the manner of people's infecting one another; namely, that it was not the sick people only from whom the plague was immediately received by others that were sound, but the well. To explain myself: by the sick people I mean those who were known to be sick, had taken their beds, had been under cure, or had swellings and tumours upon them, and the like; these everybody could beware of; they were either in their beds or in such condition as could not be concealed.

By the well I mean such as had received the contagion, and had it really upon them, and in their blood, yet did not show the consequences of it in their countenances: nay, even were not sensible of it themselves, as many were not for several days. These breathed death in every place, and upon everybody who came near them; nay, their very clothes retained the infection, their hands would infect the things they touched, especially if they were warm and sweaty, and they were generally apt to sweat too.

Now it was impossible to know these people, nor did they sometimes, as I have said, know themselves to be infected. These were the people that so often dropped down and fainted in the streets; for oftentimes they would go about the streets to the last, till on a sudden they would sweat, grow faint, sit down at a door and die. It is true, finding themselves thus, they would struggle hard to get home to their own doors, or at other times would be just able to go into their houses and die instantly; other times they would go about till they had the very tokens come out upon them, and yet not know it, and would die in an hour or two after they came home, but be well as long as they were abroad. These were the dangerous people; these were the people of whom the well people ought to have been afraid; but then, on the other side, it was impossible to know them.

And this is the reason why it is impossible in a visitation to prevent the spreading of the plague by the utmost human vigilance: viz., that it is impossible to know the infected people from the sound, or that the infected people should perfectly know themselves. I knew a man who conversed freely in London all the season of the plague in 1665, and kept about him an antidote or cordial on purpose to take when he thought himself in any danger, and he had such a rule to know or have warning of the danger by as indeed I never met with before or since. How far it may be depended on I know not. He had a wound in his leg, and whenever he came among any people that were not sound, and the infection began to affect him, he said he could know it by that signal, viz., that his wound in his leg would smart, and look pale and white; so as soon as ever he felt it smart it was time for him to withdraw, or to take care of himself, taking his drink, which he always carried about him for that purpose. Now it seems he found his wound would smart many times when he was in company with such who thought themselves to be sound, and who appeared so to one another; but he would presently rise up and say publicly, 'Friends, here is somebody in the room that has the plague', and so would immediately break up the company. This was indeed a faithful monitor to all people that the plague is not to be avoided by those that converse promiscuously in a town infected, and people have it when they know it not, and that they likewise give it to others when they know not that they have it themselves; and in this case shutting up the well or removing the sick will not do it, unless they can go back and shut up all those that the sick had conversed with, even before they knew themselves to be sick, and none knows how far to carry that back, or where to stop; for none knows when or where or how they may have received the infection, or from whom.

This I take to be the reason which makes so many people talk of the air being corrupted and infected, and that they need not be cautious of whom they converse with, for that the contagion was in the air. I have seen them in strange agitations and surprises on this account. 'I have never come near any infected body', says the disturbed person; 'I have conversed with none but sound, healthy people, and yet I have gotten the distemper!' 'I am sure I am struck from Heaven', says another, and he falls to the serious part. Again, the first goes on exclaiming, 'I have come near no infection or any infected person; I am sure it is the air. We draw in death when we breathe, and therefore 'tis the hand of God; there is no withstanding it.' And this at last made many people, being hardened to the danger, grow less concerned at it; and less cautious towards the latter end of the time, and when it was come to its height, than they were at first. Then, with a kind of a Turkish predestinarianism, they would say, if it pleased God to strike them, it was all one whether they went abroad or stayed at home; they could not escape it, and therefore they went boldly about, even into infected houses and infected company; visited sick people; and, in short, lay in the beds with their wives or relations when they were infected. And what was the consequence, but the same that is the consequence in Turkey, and in those countries where they do those things—namely, that they were infected too, and died by hundreds and thousands?

I would be far from lessening the awe of the judgements of God and the reverence to His providence which ought always to be on our minds on such occasions as these. Doubtless the visitation itself is a stroke from Heaven upon a city, or country, or nation where it falls; a messenger of His vengeance, and a loud call to that nation or country or city to humiliation and repentance, according to that of the prophet Jeremiah (xviii. 7, 8): 'At what instant I shall speak concerning a nation, and concerning a kingdom, to pluck up, and to pull down, and to destroy it; if that nation against whom I have pronounced turn from their evil, I will repent of the evil that I thought to do unto them.' Now to prompt due impressions of the awe of God on the minds of men on such occasions, and not to lessen them, it is that I have left those minutes upon record.

I say, therefore, I reflect upon no man for putting the reason of those things upon the immediate hand of God, and the appointment and direction of His providence; nay, on the contrary, there were many wonderful deliverances of persons from infection, and deliverances of persons when infected, which intimate singular and remarkable providence in the particular instances to which they refer; and I esteem my own deliverance to be one next to miraculous, and do record it with thankfulness.

But when I am speaking of the plague as a distemper arising from natural causes, we must consider it as it was really propagated by natural means; nor is it at all the less a judgement for its being under the conduct of human causes and effects; for, as the Divine Power has formed the whole scheme of nature and maintains nature in its course, so the same Power thinks fit to let His own actings with men, whether of mercy or judgement, to go on in the ordinary course of natural causes; and He is pleased to act by those natural causes as the ordinary means, excepting and reserving to Himself nevertheless a power to act in a supernatural way when He sees occasion. Now 'tis evident that in the case of an infection there is no apparent extraordinary occasion for supernatural operation, but the ordinary course of things appears sufficiently armed, and made capable of all the effects that Heaven usually directs by a contagion. Among these causes and effects, this of the secret conveyance of infection, imperceptible and unavoidable, is more than sufficient to execute the fierceness of Divine vengeance, without putting it upon supernaturals and miracle.

The acute penetrating nature of the disease itself was such, and the infection was received so imperceptibly, that the most exact caution could not secure us while in the place. But I must be allowed to believe—and I have so many examples fresh in my memory to convince me of it, that I think none can resist their evidence—I say, I must be allowed to believe that no one in this whole nation ever received the sickness or infection but who received it in the ordinary way of infection from somebody, or the clothes or touch or stench of somebody that was infected before.

The manner of its coming first to London proves this also, viz., by goods brought over from Holland, and brought thither from the Levant; the first breaking of it out in a house in Long Acre where those goods were carried and first opened; its spreading from that house to other houses by the visible unwary conversing with those who were sick; and the infecting the parish officers who were employed about the persons dead, and the like. These are known authorities for this great foundation point—that it went on and proceeded from person to person and from house to house, and no otherwise. In the first house that was infected there died four persons. A neighbour, hearing the mistress of the first house was sick, went to visit her, and went home and gave the distemper to her family, and died, and all her household. A minister, called to pray with the first sick person in the second house, was said to sicken immediately and die with several more in his house. Then the physicians began to consider, for they did not at first dream of a general contagion. But the physicians being sent to inspect the bodies, they assured the people that it was neither more or less than the plague, with all its terrifying particulars, and that it threatened an universal infection, so many people having already conversed with the sick or distempered, and having, as might be supposed, received infection from them, that it would be impossible to put a stop to it.

Here the opinion of the physicians agreed with my observation afterwards, namely, that the danger was spreading insensibly, for the sick could infect none but those that came within reach of the sick person; but that one man who may have really received the infection and knows it not, but goes abroad and about as a sound person, may give the plague to a thousand people, and they to greater numbers in proportion, and neither the person giving the infection or the persons receiving it know anything of it, and perhaps not feel the effects of it for several days after.

For example, many persons in the time of this visitation never perceived that they were infected till they found to their unspeakable surprise, the tokens come out upon them; after which they seldom lived six hours; for those spots they called the tokens were really gangrene spots, or mortified flesh in small knobs as broad as a little silver penny, and hard as a piece of callus or horn; so that, when the disease was come up to that length, there was nothing could follow but certain death; and yet, as I said, they knew nothing of their being infected, nor found themselves so much as out of order, till those mortal marks were upon them. But everybody must allow that they were infected in a high degree before. And must have been so some time, and consequently their breath, their sweat, their very clothes, were contagious for many days before.

This occasioned a vast variety of cases which physicians would have much more opportunity to remember than I; but some came within the compass of my observation or hearing, of which I shall name a few.

A certain citizen who had lived safe and untouched till the month of September, when the weight of the distemper lay more in the city than it had done before, was mighty cheerful, and something too bold (as I think it was) in his talk of how secure he was, how cautious he had been, and how he had never come near any sick body. Says another citizen, a neighbour of his, to him one day, 'Do not be too confident, Mr—; it is hard to say who is sick and who is well, for we see men alive and well to outward appearance one hour, and dead the next.' 'That is true', says the first man, for he was not a man presumptuously secure, but had escaped a long while—and men, as I said above, especially in the city began to be over-easy upon that score. That is true,' says he; 'I do not think myself secure, but I hope I have not been in company with any person that there has been any danger in.' 'No?' says his neighbour. 'Was not you at the Bull Head Tavern in Gracechurch Street with Mr—the night before last?' 'Yes,' says the first, 'I was; but there was nobody there that we had any reason to think dangerous.' Upon which his neighbour said no more, being unwilling to surprise him; but this made him more inquisitive, and as his neighbour appeared backward, he was the more impatient, and in a kind of warmth says he aloud, 'Why, he is not dead, is he?' Upon which his neighbour still was silent, but cast up his eyes and said something to himself; at which the first citizen turned pale, and said no more but this, 'Then I am a dead man too', and went home immediately and sent for a neighbouring apothecary to give him something preventive, for he had not yet found himself ill; but the apothecary, opening his breast, fetched a sigh, and said no more but this, 'Look up to God'; and the man died in a few hours.

Now let any man judge from a case like this if it is possible for the regulations of magistrates, either by shutting up the sick or removing them, to stop an infection which spreads itself from man to man even while they are perfectly well and insensible of its approach, and may be so for many days.

It may be proper to ask here how long it may be supposed men might have the seeds of the contagion in them before it discovered itself in this fatal manner, and how long they might go about seemingly whole, and yet be contagious to all those that came near them. I believe the most experienced physicians cannot answer this question directly any more than I can; and something an ordinary observer may take notice of, which may pass their observations. The opinion of physicians abroad seems to be that it may lie dormant in the spirits or in the blood-vessels a very considerable time. Why else do they exact a quarantine of those who came into their harbours and ports from suspected places? Forty days is, one would think, too long for nature to struggle with such an enemy as this, and not conquer it or yield to it. But I could not think, by my own observation, that they can be infected so as to be contagious to others above fifteen or sixteen days at furthest; and on that score it was, that when a house was shut up in the city and any one had died of the plague, but nobody appeared to be ill in the family for sixteen or eighteen days after, they were not so strict but that they would connive at their going privately abroad; nor would people be much afraid of them afterward, but rather think they were fortified the better, having not been vulnerable when the enemy was in their own house; but we sometimes found it had lain much longer concealed.

Upon the foot of all these observations I must say that though Providence seemed to direct my conduct to be otherwise, yet it is my opinion, and I must leave it as a prescription, viz., that the best physic against the plague is to run away from it. I know people encourage themselves by saying God is able to keep us in the midst of danger, and able to overtake us when we think ourselves out of danger; and this kept thousands in the town whose carcasses went into the great pits by cartloads, and who, if they had fled from the danger, had, I believe, been safe from the disaster; at least 'tis probable they had been safe.

And were this very fundamental only duly considered by the people on any future occasion of this or the like nature, I am persuaded it would put them upon quite different measures for managing the people from those that they took in 1665, or than any that have been taken abroad that I have heard of. In a word, they would consider of separating the people into smaller bodies, and removing them in time farther from one another—and not let such a contagion as this, which is indeed chiefly dangerous to collected bodies of people, find a million of people in a body together, as was very near the case before, and would certainly be the case if it should ever appear again.

The plague, like a great fire, if a few houses only are contiguous where it happens, can only bum a few houses; or if it begins in a single, or, as we call it, a lone house, can only bum that lone house where it begins. But if it begins in a close-built town or city and gets a head, there its fury increases: it rages over the whole place, and consumes all it can reach.

I could propose many schemes on the foot of which the government of this city, if ever they should be under the apprehensions of such another enemy (God forbid they should), might ease themselves of the greatest part of the dangerous people that belong to them; I mean such as the begging, starving, labouring poor, and among them chiefly those who, in case of a siege, are called the useless mouths; who being then pmdently and to their own advantage disposed of, and the wealthy inhabitants disposing of themselves and of their servants and children, the city and its adjacent parts would be so effectually evacuated that there would not be above a tenth part of its people left together for the disease to take hold upon. But suppose them to be a fifth part, and that two hundred and fifty thousand people were left: and if it did seize upon them, they would, by their living so much at large, be much better prepared to defend themselves against the infection, and be less liable to the effects of it than if the same number of people lived dose together in one smaller city such as Dublin or Amsterdam or the like.

It is tme hundreds, yea, thousands of families fled away at this last plague, but then of them, many fled too late, and not only died in their flight, but carried the distemper with them into the countries where they went and infected those whom they went among for safety; which confounded the thing, and made that be a propagation of the distemper which was the best means to prevent it; and this too is an evidence of it, and brings me back to what I only hinted at before, but must speak more fully to here, namely, that men went about apparently well many days after they had the taint of the disease in their vitals, and after their spirits were so seized as that they could never escape it, and that all the while they did so they were dangerous to others; I say, this proves that so it was; for such people infected the very towns they went through, as well as the families they went among; and it was by that means that almost all the great towns in England had the distemper among them, more or less, and always they would tell you such a Londoner or such a Londoner brought it down.

It must not be omitted that when I speak of those people who were really thus dangerous, I suppose them to be utterly ignorant of their own conditions; for if they really knew their circumstances to be such as indeed they were, they must have been a kind of wilful murtherers if they would have gone abroad among healthy people—and it would have verified indeed the suggestion which I mentioned above, and which I thought seemed untrue: viz., that the infected people were utterly careless as to giving the infection to others, and rather forward to do it than not; and I believe it was partly from this very thing that they raised that suggestion, which I hope was not really true in fact.

I confess no particular case is sufficient to prove a general, but I could name several people within the knowledge of some of their neighbours and families yet living who showed the contrary to an extreme. One man, a master of a family in my neighbourhood, having had the distemper, he thought he had it given him by a poor workman whom he employed, and whom he went to his house to see, or went for some work that he wanted to have finished; and he had some apprehensions even while he was at the poor workman's door, but did not discover it fully; but the next day it discovered itself, and he was taken very in, upon which he immediately caused himself to be carried into an outbuilding which he had in his yard, and where there was a chamber over a workhouse (the man being a brazier). Here he lay, and here he died, and would be tended by none of his neighbours, but by a nurse from abroad; and would not suffer his wife, nor children, nor servants to come up into the room, lest they should be infected—but sent them his blessing and prayers for them by the nurse, who spoke it to them at a distance, and all this for fear of giving them the distemper; and without which he knew, as they were kept up, they could not have it.

And here I must observe also that the plague, as I suppose all distempers do, operated in a different manner on differing constitutions; some were immediately overwhelmed with it, and it came to violent fevers, vomitings, insufferable headaches, pains in the back, and so up to ravings and ragings with those pains; others with swellings and tumours in the neck or groin, or armpits, which till they could be broke put them into insufferable agonies and torment; while others, as I have observed, were silently infected, the fever preying upon their spirits insensibly, and they seeing little of it till they fell into swooning, and faintings, and death without pain. I am not physician enough to enter into the particular reasons and manner of these differing effects of one and the same distemper, and of its differing operation in several bodies; nor is it my business here to record the observations which I really made, because the doctors themselves have done that part much more effectually than I can do, and because my opinion may in some things differ from theirs. I am only relating what I know, or have heard, or believe of the particular cases, and what fell within the compass of my view, and the different nature of the infection as it appeared in the particular cases which I have related; but this may be added too: that though the former sort of those cases, namely, those openly visited, were the worst for themselves as to pain—I mean those that had such fevers, vomitings, headaches, pains, and swellings, because they died in such a dreadful manner—yet the latter had the worst state of the disease; for in the former they frequently recovered, especially if the swellings broke; but the latter was inevitable death; no cure, no hell), could be possible, nothing could follow but death. And it was worse also to others, because, as above, it secretly and unperceived by others or by themselves, communicated death to those they conversed with, 'the penetrating poison insinuating itself into their blood in a manner which it is impossible to describe, or indeed conceive.

This infecting and being infected without so much as its being known to either person is evident from two sorts of cases which frequently happened at that time; and there is hardly anybody living who was in London during the infection but must have known several of the cases of both sorts.

(1) Fathers and mothers have gone about as if they had been well, and have believed themselves to be so, till they have insensibly infected and been the destruction of their whole families, which they would have been far from doing if they had the least

apprehensions of their being unsound and dangerous themselves. A family, whose story I have heard, was thus infected by the father; and the distemper began to appear upon some of them even before he found it upon himself. But searching more narrowly, it appeared he had been affected some time; and as soon as he found that his family had been poisoned by himself he went distracted, and would have laid violent hands upon himself, but was kept from that by those who looked to him, and in a few days died.

(2) The other particular is, that many people having been well to the best of their own judgement, or by the best observation which they could make of themselves for several days, and only finding a decay of appetite, or a light sickness upon their stomachs; nay, some whose appetite has been strong, and even craving, and only a light pain in their heads, have sent for physicians to know what ailed them, and have been found, to their great surprise, at the brink of death: the tokens upon them, or the plague grown up to an incurable height.

It was very sad to reflect how such a person as this last mentioned above had been a walking destroyer perhaps for a week or a fortnight before that; how he had ruined those that he would have hazarded his life to save, and had been breathing death upon them, even perhaps in his tender kissing and embracings of his own children. Yet thus certainly it was, and often has been, and I could give many particular cases where it has been so. If then the blow is thus insensibly striking—if the arrow flies thus unseen, and cannot be discovered—to what purpose are all the schemes for shutting up or removing the sick people? Those schemes cannot take place but upon those that appear to be sick, or to be infected; whereas there are among them at the same time thousands of people who seem to be well, but are all that while carrying death with them into all companies which they come into.

This frequently puzzled our physicians, and especially the apothecaries and surgeons, who knew not how to discover the sick from the sound; they all allowed that it was really so, that many people had the plague in their very blood, and preying upon their spirits, and were in themselves but walking putrefied carcasses whose breath was infectious and their sweat poison, and yet were as well to look on as other people, and even knew it not themselves; I say, they all allowed that it was really true in fact, but they knew not how to propose a discovery.

My friend Dr Heath was of opinion that it might be known by the smell of their breath; but then, as he said, who durst smell to that breath for his information? since, to know it, he must draw the stench of the plague up into his own brain, in order to distinguish the smell! I have heard it was the opinion of others that it might be distinguished by the party's breathing upon a piece of glass, where, the breath condensing, there might living creatures be seen by a microscope, of strange, monstrous, and frightful shapes, such as dragons, snakes, serpents, and devils, horrible to behold. But this I very much question the truth of, and we had no microscopes at that time, as I remember, to make the experiment with.

It was the opinion also of another learned man, that the breath of such a person would poison and instantly kill a bird; not only a small bird, but even a cock or hen, and that, if it did not immediately kill the latter, it would cause them to be roupy, as they call it; particularly that if they had laid any eggs at any time, they would be all rotten. But those are opinions which I never found supported by any experiments, or heard of others that had seen it; so I leave them as I find them; only with this remark, namely, that I think the probabilities are very strong for them.

Some have proposed that such persons should breathe hard upon warm water, and that they would leave an unusual scum upon it, or upon several other things, especially such as are of a glutinous substance and are apt to receive a scum and support it.

But from the whole I found that the nature of this contagion was such that it was impossible to discover it at all, or to prevent its spreading from one to another by any human skill.

Here was indeed one difficulty which I could never thoroughly get over to this time, and which there is but one way of answering that I know of, and it is this, viz., the first person that died of the plague was on December 20, or thereabouts, 1664, and in or about long Acre; whence the first person had the infection was generally said to be from a parcel of silks imported from Holland, and first opened in that house.

But after this we heard no more of any person dying of the plague, or of the distemper being in that place, till the 9th of February, which was about seven weeks after, and then one more was buried out of the same house. Then it was hushed, and we were perfectly easy as to the public for a great while; for there were no more entered in the weekly bill to be dead of the plague till the 22nd of April, when there was two more buried, not out of the same house, but out of the same street; and, as near as I can remember, it was out of the next house to the first. This was nine weeks asunder, and after this we had no more till a fortnight, and then it broke out in several streets and spread every way. Now the question seems to lie thus: Where lay the seeds of the infection all this while? How came it to stop so long, and not stop any longer? Either the distemper did not come immediately by contagion from body to body, or, if it did, then a body may be capable to continue infected without the disease discovering itself many days, nay, weeks together; even not a quarantine of days only, but soixantine; not only forty days, but sixty days or longer.

It is true there was, as I observed at first, and is well known to many yet living, a very cold winter and a long frost which continued three months; and this, the doctors say, might check the infection; but then the learned must allow me to say that if, according to their notion, the disease was (as I may say) only frozen up, it would like a frozen river have returned to its usual force and current when it thawed—whereas the principal recess of this infection, which was from February to April, was after the frost was broken and the weather mild and warm.

But there is another way of solving all this difficulty, which I think my own remembrance of the thing will supply; and that is, the fact is not granted—namely, that there died none in those long intervals, viz., from the 20th of December to the 9th of February, and from thence to the 22nd of April. The weekly bills are the only evidence on the other side, and those bills were not of credit enough, at least with me, to support an hypothesis or determine a question of such importance as this; for it was our received opinion at that time, and I believe upon very good grounds, that the fraud lay in the parish officers, searchers, and persons appointed to give account of the dead, and what diseases they died of; and as people were very loath at first to have the neighbours believe their houses were infected, so they gave money to procure, or otherwise procured, the dead persons to be returned as dying of other distempers; and this I know was practised afterwards in many places, I believe I might say in all places where the distemper came, as will be seen by the vast increase of the numbers placed in the weekly bills under other articles of diseases during the time of the infection. For example, in the months of July and August, when the plague was coming on to its highest pitch, it was very ordinary to have from a thousand to twelve hundred, nay, to almost fifteen hundred a week of other distempers. Not that the numbers of those distempers were really increased to such a degree, but the great number of families and houses where really the infection was, obtained the favour to have their dead be returned of other distempers, to prevent the shutting up their houses. For example:—

Dead of other diseases beside the plague:—
From the 18th July to the 25th 942
" 25th July " 1st August 1004
" 1st August" 8th 1213
" 8th " 15th 1439
" 15th " 22nd 1331
" 22nd " 29th 1394
" 29th " 5th September 1264
" 5th September to the 12th 1056
" 12th " 19th 1132
" 19th " 26th 927

Now it was not doubted but the greatest part of these, or a great part of them, were dead of the plague, but the officers were prevailed with to return them as above, and the numbers of some particular articles of distempers discovered is as follows:—

Aug. Aug. Aug. Aug. Aug. Sept. Sept. Sept.
1 8 15 22 29 5 12 19
to 8 to 15 to 22 to 29 to Sept.5 to 12 to 19 to 26
Fever 314 353 348 383 364 332 309 268
Spotted 174 190 166 165 157 97 101 65
Fever
Surfeit 85 87 74 99 68 45 49 36
Teeth 90 113 111 133 138 128 121 112
663 743 699 780 727 602 580 481

There were several other articles which bore a proportion to these, and which, it is easy to perceive, were increased on the same account, as aged, consumptions, vomitings, imposthumes, gripes, and the like, many of which were not doubted to be infected people; but as it was of the utmost consequence to families not to be known to be infected, if it was possible to avoid it, so they took all the measures they could to have it not believed, and if any died in their houses, to get them returned to the examiners, and by the searchers, as having died of other distempers.

This, I say, will account for the long interval which, as I have said, was between the dying of the first persons that were returned in the bill to be dead of the plague and the time when the distemper spread openly and could not be concealed.

Besides, the weekly bills themselves at that time evidently discover the truth; for, while there was no mention of the plague, and no increase after it had been mentioned, yet it was apparent that there was an increase of those distempers which bordered nearest upon it; for example, there were eight, twelve, seventeen of the spotted fever in a week, when there were none, or but very few, of the plague; whereas before, one, three, or four were the ordinary weekly numbers of that distemper. Likewise, as I observed before, the burials increased weekly in that particular parish and the parishes adjacent more than in any other parish, although there were none set down of the plague; all which tells us, that the infection was handed on, and the succession of the distemper really preserved, though it seemed to us at that time to be ceased, and to come again in a manner surprising.

It might be, also, that the infection might remain in other parts of the same parcel of goods which at first it came in, and which might not be perhaps opened, or at least not fully, or in the clothes of the first infected person; for I cannot think that anybody could be seized with the contagion in a fatal and mortal degree for nine weeks together, and support his state of health so well as even not to discover it to themselves; yet if it were so, the argument is the stronger in favour of what I am saying: namely, that the infection is retained in bodies apparently well, and conveyed from them to those they converse with, while it is known to neither the one nor the other.

Great were the confusions at that time upon this very account, and when people began to be convinced that the infection was received in this surprising manner from persons apparently well, they began to be exceeding shy and jealous of every one that came near them. Once, on a public day, whether a Sabbath-day or not I do not remember, in Aldgate Church, in a pew full of people, on a sudden one fancied she smelt an ill smell. Immediately she fancies the plague was in the pew, whispers her notion or suspicion to the next, then rises and goes out of the pew. It immediately took with the next, and so to them all; and every one of them, and of the two or three adjoining pews, got up and went out of the church, nobody knowing what it was offended them, or from whom.

This immediately filled everybody's mouths with one preparation or other, such as the old woman directed, and some perhaps as physicians directed, in order to prevent infection by the breath of others; insomuch that if we came to go into a church when it was anything full of people, there would be such a mixture of smells at the entrance that it was much more strong, though perhaps not so wholesome, than if you were going into an apothecary's or druggist's shop. In a word, the whole church was like a smelling-bottle; in one corner it was all perfumes; in another, aromatics, balsamics, and variety of drugs and herbs; in another, salts and spirits, as every one was furnished for their own preservation. Yet I observed that after people were possessed, as I have said, with the belief, or rather assurance, of the infection being thus carried on by persons apparently in health, the churches and meeting-houses were much thinner of people than at other times before that they used to be. For this is to be said of the people of London, that during the whole time of the pestilence the churches or meetings were never wholly shut up, nor did the people decline coming out to the public worship of God, except only in some parishes when the violence of the distemper was more particularly in that parish at that time, and even then no longer than it continued to be so.

Indeed nothing was more strange than to see with what courage the people went to the public service of God, even at that time when they were afraid to stir out of their own houses upon any other occasion; this, I mean, before the time of desperation, which I have mentioned already. This was a proof of the exceeding populousness of the city at the time of the infection, notwithstanding the great numbers that were gone into the country at the first alarm, and that fled out into the forests and woods when they were further terrified with the extraordinary increase of it. For when we came to see the crowds and throngs of people which appeared on the Sabbath-days at the churches, and especially in those parts of the town where the plague was abated, or where it was not yet come to its height, it was amazing. But of this I shall speak again presently. I return in the meantime to the article of infecting one another at first, before people came to right notions of the infection, and of infecting one another. People were only shy of those that were really sick, a man with a cap upon his head, or with clothes round his neck, which was the case of those that had swellings there. Such was indeed frightful; but when we saw a gentleman dressed, with his band on and his gloves in his hand, his hat upon his head, and his hair combed, of such we bad not the least apprehensions, and people conversed a great while freely, especially with their neighbours and such as they knew. But when the physicians assured us that the danger was as well from the sound (that is, the seemingly sound) as the sick, and that those people who thought themselves entirely free were oftentimes the most fatal, and that it came to be generally understood that people were sensible of it, and of the reason of it; then, I say, they began to be jealous of everybody, and a vast number of people locked themselves up, so as not to come abroad into any company at all, nor suffer any that had been abroad in promiscuous company to come into their houses, or near them—at least not so near them as to be within the reach of their breath or of any smell from them; and when they were obliged to converse at a distance with strangers, they would always have preservatives in their mouths and about their clothes to repel and keep off the infection.

It must be acknowledged that when people began to use these cautions they were less exposed to danger, and the infection did not break into such houses so furiously as it did into others before; and thousands of families were preserved (speaking with due reserve to the direction of Divine Providence) by that means.

But it was impossible to beat anything into the heads of the poor. They went on with the usual impetuosity of their tempers, full of outcries and lamentations when taken, but madly careless of themselves, foolhardy and obstinate, while they were well. Where they could get employment they pushed into any kind of business, the most dangerous and the most liable to infection; and if they were spoken to, their answer would be, 'I must trust to God for that; if I am taken, then I am provided for, and there is an end of me', and the like. Or thus, 'Why, what must I do? I can't starve. I had as good have the plague as perish for want. I have no work; what could I do? I must do this or beg.' Suppose it was burying the dead, or attending the sick, or watching infected houses, which were all terrible hazards; but their tale was generally the same. It is true, necessity was a very justifiable, warrantable plea, and nothing could be better; but their way of talk was much the same where the necessities were not the same. This adventurous conduct of the poor was that which brought the plague among them in a most furious manner; and this, joined to the distress of their circumstances when taken, was the reason why they died so by heaps; for I cannot say I could observe one jot of better husbandry among them, I mean the labouring poor, while they were all well and getting money than there was before, but as lavish, as extravagant, and as thoughtless for tomorrow as ever; so that when they came to be taken sick they were immediately in the utmost distress, as well for want as for sickness, as well for lack of food as lack of health.

This misery of the poor I had many occasions to be an eyewitness of, and sometimes also of the charitable assistance that some pious people daily gave to such, sending them relief and supplies both of food, physic, and other help, as they found they wanted; and indeed it is a debt of justice due to the temper of the people of that day to take notice here, that not only great sums, very great sums of money were charitably sent to the Lord Mayor and aldermen for the assistance and support of the poor distempered people, but abundance of private people daily distributed large sums of money for their relief, and sent people about to inquire into the condition of particular distressed and visited families, and relieved them; nay, some pious ladies were so transported with zeal in so good a work, and so confident in the protection of Providence in discharge of the great duty of charity, that they went about in person distributing alms to the poor, and even visiting poor families, though sick and infected, in their very houses, appointing nurses to attend those that wanted attending, and ordering apothecaries and surgeons, the first to supply them with drugs or plasters, and such things as they wanted; and the last to lance and dress the swellings and tumours, where such were wanting; giving their blessing to the poor in substantial relief to them, as well as hearty prayers for them.

I will not undertake to say, as some do, that none of those charitable people were suffered to fall under the calamity itself; but this I may say, that I never knew any one of them that miscarried, which I mention for the encouragement of others in case of the like distress; and doubtless, if they that give to the poor lend to the Lord, and He will repay them, those that hazard their lives to give to the poor, and to comfort and assist the poor in such a misery as this, may hope to be protected in the work.

Nor was this charity so extraordinary eminent only in a few, but (for I cannot lightly quit this point) the charity of the rich, as well in the city and suburbs as from the country, was so great that, in a word, a prodigious number of people who must otherwise inevitably have perished for want as well as sickness were supported and subsisted by it; and though I could never, nor I believe any one else, come to a full knowledge of what was so contributed, yet I do believe that, as I heard one say that was a critical observer of that part, there was not only many thousand pounds contributed, but many hundred thousand pounds, to the relief of the poor of this distressed, afflicted city; nay, one man affirmed to me that he could reckon up above one hundred thousand pounds a week, which was distributed by the churchwardens at the several parish vestries by the Lord Mayor and aldermen in the several wards and precincts, and by the particular direction of the court and of the justices respectively in the parts where they resided, over and above the private charity distributed by pious bands in the manner I speak of; and this continued for many weeks together.

I confess this is a very great sum; but if it be true that there was distributed in the parish of Cripplegate only, 17,800 in one week to the relief of the poor, as I heard reported, and which I really believe was true, the other may not be improbable.

It was doubtless to be reckoned among the many signal good providences which attended this great city, and of which there were many other worth recording—I say, this was a very remarkable one, that it pleased God thus to move the hearts of the people in all parts of the kingdom so cheerfully to contribute to the relief and support of the poor at London, the good consequences of which were felt many ways, and particularly in preserving the lives and recovering the health of so many thousands, and keeping so many thousands of families from perishing and starving.

And now I am talking of the merciful disposition of Providence in this time of calamity, I cannot but mention again, though I have spoken several times of it already on other accounts, I mean that of the progression of the distemper; how it began at one end of the town, and proceeded gradually and slowly from one part to another, and like a dark cloud that passes over our heads, which, as it thickens and overcasts the air at one end, clears up at the other end; so, while the plague went on raging from west to east, as it went forwards east, it abated in the west, by which means those parts of the town which were not seized, or who were left, and where it had spent its fury, were (as it were) spared to help and assist the other; whereas, had the distemper spread itself over the whole city and suburbs, at once, raging in all places alike, as it has done since in some places abroad, the whole body of the people must have been overwhelmed, and there would have died twenty thousand a day, as they say there did at Naples;, nor would the people have been able to have helped or assisted one another.

For it must be observed that where the plague was in its full force, there indeed the people were very miserable, and the consternation was inexpressible. But a little before it reached even to that place, or presently after it was gone, they were quite another sort of people; and I cannot but acknowledge that there was too much of that common temper of mankind to be found among us all at that time, namely, to forget the deliverance when the danger is past. But I shall come to speak of that part again.

Part 3

It must not be forgot here to take some notice of the state of trade during the time of this common calamity, and this with respect to foreign trade, as also to our home trade.

As to foreign trade, there needs little to be said. The trading nations of Europe were all afraid of us; no port of France, or Holland, or Spain, or Italy would admit our ships or correspond with us; indeed we stood on ill terms with the Dutch, and were in a furious war with them, but though in a bad condition to fight abroad, who had such dreadful enemies to struggle with at home.

Our merchants were accordingly at a full stop; their ships could go nowhere—that is to say, to no place abroad; their manufactures and merchandise—that is to say, of our growth—would not be touched abroad. They were as much afraid of our goods as they were of our people; and indeed they had reason: for our woollen manufactures are as retentive of infection as human bodies, and if packed up by persons infected, would receive the infection and be as dangerous to touch as a man would be that was infected; and therefore, when any English vessel arrived in foreign countries, if they did take the goods on shore, they always caused the bales to be opened and aired in places appointed for that purpose. But from London they would not suffer them to come into port, much less to unlade their goods, upon any terms whatever, and this strictness was especially used with them in Spain and Italy. In Turkey and the islands of the Arches indeed, as they are called, as well those belonging to the Turks as to the Venetians, they were not so very rigid. In the first there was no obstruction at all; and four ships which were then in the river loading for Italy—that is, for Leghorn and Naples—being denied product, as they call it, went on to Turkey, and were freely admitted to unlade their cargo without any difficulty; only that when they arrived there, some of their cargo was not fit for sale in that country; and other parts of it being consigned to merchants at Leghorn, the captains of the ships had no right nor any orders to dispose of the goods; so that great inconveniences followed to the merchants. But this was nothing but what the necessity of affairs required, and the merchants at Leghorn and Naples having notice given them, sent again from thence to take care of the effects which were particularly consigned to those ports, and to bring back in other ships such as were improper for the markets at Smyrna and Scanderoon.

The inconveniences in Spain and Portugal were still greater, for they would by no means suffer our ships, especially those from London, to come into any of their ports, much less to unlade. There was a report that one of our ships having by stealth delivered her cargo, among which was some bales of English cloth, cotton, kerseys, and such-like goods, the Spaniards caused all the goods to be burned, and punished the men with death who were concerned in carrying them on shore. This, I believe, was in part true, though I do not affirm it; but it is not at all unlikely, seeing the danger was really very great, the infection being so violent in London.

I heard likewise that the plague was carried into those countries by some of our ships, and particularly to the port of Faro in the kingdom of Algarve, belonging to the King of Portugal, and that several persons died of it there; but it was not confirmed.

On the other hand, though the Spaniards and Portuguese were so shy of us, it is most certain that the plague (as has been said) keeping at first much at that end of the town next Westminster, the merchandising part of the town (such as the city and the water-side) was perfectly sound till at least the beginning of July, and the ships in the river till the beginning of August; for to the 1st of July there had died but seven within the whole city, and but sixty within the liberties, but one in all the parishes of Stepney, Aldgate, and Whitechappel, and but two in the eight parishes of Southwark. But it was the same thing abroad, for the bad news was gone over the whole world that the city of London was infected with the plague, and there was no inquiring there how the infection proceeded, or at which part of the town it was begun or was reached to.

Besides, after it began to spread it increased so fast, and the bills grew so high all on a sudden, that it was to no purpose to lessen the report of it, or endeavour to make the people abroad think it better than it was; the account which the weeMy bills gave in was sufficient; and that there died two thousand to three or-four thousand a week was sufficient to alarm the whole trading part of the world; and the following time, being so dreadful also in the very city itself, put the whole world, I say, upon their guard against it.

You may be sure, also, that the report of these things lost nothing in the carriage. The plague was itself very terrible, and the distress of the people very great, as you may observe of what I have said. But the rumour was infinitely greater, and it must not be wondered that our friends abroad (as my brother's correspondents in particular were told there, namely, in Portugal and Italy, where he chiefly traded) [said] that in London there died twenty thousand in a week; that the dead bodies lay unburied by heaps; that the living were not sufficient to bury the dead or the sound to look after the sick; that all the kingdom was infected likewise, so that it was an universal malady such as was never heard of in those parts of the world; and they could hardly beheve us when we gave them an account how things really were, and how there was not above one-tenth part of the people dead; that there was 500,000, left that lived all the time in the town; that now the people began to walk the streets again, and those who were fled to return, there was no miss of the usual throng of people in the streets, except as every family might miss their relations and neighbours, and the like. I say they could not believe these things; and if inquiry were now to be made in Naples, or in other cities on the coast of Italy, they would tell you that there was a dreadful infection in London so many years ago, in which, as above, there died twenty thousand in a week, &c., just as we have had it reported in London that there was a plague in the city of Naples in the year 1656, in which there died 20,000 people in a day, of which I have had very good satisfaction that it was utterly false.

But these extravagant reports were very prejudicial to our trade, as well as unjust and injurious in themselves, for it was a long time after the plague was quite over before our trade could recover itself in those parts of the world; and the Flemings and Dutch (but especially the last) made very great advantages of it, having all the market to themselves, and even buying our manufactures in several parts of England where the plague was not, and carrying them to Holland and Flanders, and from thence transporting them to Spain and to Italy as if they had been of their own making.

But they were detected sometimes and punished: that is to say, their goods confiscated and ships also; for if it was true that our manufactures as well as our people were infected, and that it was dangerous to touch or to open and receive the smell of them, then those people ran the hazard by that clandestine trade not only of carrying the contagion into their own country, but also of infecting the nations to whom they traded with those goods; which, considering how many fives might be lost in consequence of such an action, must be a trade that no men of conscience could suffer themselves to be concerned in.

I do not take upon me to say that any harm was done, I mean of that kind, by those people. But I doubt I need not make any such proviso in the case of our own country; for either by our people of London, or by the commerce which made their conversing with all sorts of people in every country and of every considerable town necessary, I say, by this means the plague was first or last spread all over the kingdom, as well in London as in all the cities and great towns, especially in the trading manufacturing towns and seaports; so that, first or last, all the considerable places in England were visited more or less, and the kingdom of Ireland in some places, but not so universally. How it fared with the people in Scotland I had no opportunity to inquire.

It is to be observed that while the plague continued so violent in London, the outports, as they are called, enjoyed a very great trade, especially to the adjacent countries and to our own plantations. For example, the towns of Colchester, Yarmouth, and Hun, on that side of England, exported to Holland and Hamburg the manufactures of the adjacent countries for several months after the trade with London was, as it were, entirely shut up; likewise the cities of Bristol and Exeter, with the port of Plymouth, had the like advantage to Spain, to the Canaries, to Guinea, and to the West Indies, and particularly to Ireland; but as the plague spread itself every way after it had been in London to such a degree as it was in August and September, so all or most of those cities and towns were infected first or last; and then trade was, as it were, under a general embargo or at a full stop—as I shall observe further when I speak of our home trade.

One thing, however, must be observed: that as to ships coming in from abroad (as many, you may be sure, did) some who were out in all parts of the world a considerable while before, and some who when they went out knew nothing of an infection, or at least of one so terrible—these came up the river boldly, and delivered their cargoes as they were obliged to do, except just in the two months of August and September, when the weight of the infection lying, as I may say, all below Bridge, nobody durst appear in business for a while. But as this continued but for a few weeks, the homeward-bound ships, especially such whose cargoes were not liable to spoil, came to an anchor for a time short of the Pool,* or fresh-water part of the river, even as low as the river Medway, where several of them ran in; and others lay at the Nore, and in the Hope below Gravesend. So that by the latter end of October there was a very great fleet of homeward-bound ships to come up, such as the like had not been known for many years.

[*That part of the river where the ships lie up when they come home is called the Pool, and takes in all the river on both sides of the water, from the Tower to Cuckold's Point and Limehouse.]

Two particular trades were carried on by water-carriage all the while of the infection, and that with little or no interruption, very much to the advantage and comfort of the poor distressed people of the city: and those were the coasting trade for corn and the Newcastle trade for coals.

The first of these was particularly carried on by small vessels from the port of Hull and other places on the Humber, by which great quantities of corn were brought in from Yorkshire and Lincolnshire. The other part of this corn-trade was from Lynn, in Norfolk, from Wells and Burnham, and from Yarmouth, all in the same county; and the third branch was from the river Medway, and from Milton, Feversham, Margate, and Sandwich, and all the other little places and ports round the coast of Kent and Essex.

There was also a very good trade from the coast of Suffolk with corn, butter, and cheese; these vessels kept a constant course of trade, and without interruption came up to that market known still by the name of Bear Key, where they supplied the city plentifully with corn when land-carriage began to fail, and when the people began to be sick of coming from many places in the country.

This also was much of it owing to the prudence and conduct of the Lord Mayor, who took such care to keep the masters and seamen from danger when they came up, causing their corn to be bought off at any time they wanted a market (which, however, was very seldom), and causing the corn factors immediately to unlade and deliver the vessels loaden with corn, that they had very little occasion to come out of their ships or vessels, the money being always carried on board to them and put into a pail of vinegar before it was carried.

The second trade was that of coals from Newcastle-upon-Tyne, without which the city would have been greatly distressed; for not in the streets only, but in private houses and families, great quantities of coals were then burnt, even all the summer long and when the weather was hottest, which was done by the advice of the physicians. Some indeed opposed it, and insisted that to keep the houses and rooms hot was a means to propagate the temper, which was a fermentation and heat already in the blood; that it was known to spread and increase in hot weather and abate in cold; and therefore they alleged that all contagious distempers are the worse for heat, because the contagion was nourished and gained strength in hot weather, and was, as it were, propagated in heat.

Others said they granted that heat in the climate might propagate infection—as sultry, hot weather fills the air with vermin and nourishes innumerable numbers and kinds of venomous creatures which breed in our food, in the plants, and even in our bodies, by the very stench of which infection may be propagated; also that heat in the air, or heat of weather, as we ordinarily call it, makes bodies relax and faint, exhausts the spirits, opens the pores, and makes us more apt to receive infection, or any evil influence, be it from noxious pestilential vapours or any other thing in the air; but that the heat of fire, and especially of coal fires kept in our houses, or near us, had a quite different operation; the heat being not of the same kind, but quick and fierce, tending not to nourish but to consume and dissipate all those noxious fumes which the other kind of heat rather exhaled and stagnated than separated and burnt up. Besides, it was alleged that the sulphurous and nitrous particles that are often found to be in the coal, with that bituminous substance which burns, are all assisting to clear and purge the air, and render it wholesome and safe to breathe in after the noxious particles, as above, are dispersed and burnt up.

The latter opinion prevailed at that time, and, as I must confess, I think with good reason; and the experience of the citizens confirmed it, many houses which had constant fires kept in the rooms having never been infected at all; and I must join my experience to it, for I found the keeping good fires kept our rooms sweet and wholesome, and I do verily believe made our whole family so, more than would otherwise have been.

But I return to the coals as a trade. It was with no little difficulty that this trade was kept open, and particularly because, as we were in an open war with I the Dutch at that time, the Dutch capers at first took a great many of our collier-ships, which made the rest cautious, and made them to stay to come in fleets together. But after some time the capers were either afraid to take them, or their masters, the States, were afraid they should, and forbade them, lest the plague should be among them, which made them fare the better.

For the security of those northern traders, the coal-ships were ordered by my Lord Mayor not to come up into the Pool above a certain number at a time, and ordered lighters and other vessels such as the woodmongers (that is, the wharf-keepers or coal-sellers) furnished, to go down and take out the coals as low as Deptford and Greenwich, and some farther down.

Others delivered great quantities of coals in particular places where the ships could come to the shore, as at Greenwich,

Blackwall, and other places, in vast heaps, as if to be kept for sale; but were then fetched away after the ships which brought them were gone, so that the seamen had no communication with the river-men, nor so much as came near one another.

Yet all this caution could not effectually prevent the distemper getting among the colliery: that is to say among the ships, by which a great many seamen died of it; and that which was still worse was, that they carried it down to Ipswich and Yarmouth, to Newcastle-upon-Tyne, and other places on the coast—where, especially at Newcastle and at Sunderland, it carried off a great number of people.

The making so many fires, as above, did indeed consume an unusual quantity of coals; and that upon one or two stops of the ships coming up, whether by contrary weather or by the interruption of enemies I do not remember, but the price of coals was exceeding dear, even as high as 4 a chalder; but it soon abated when the ships came in, and as afterwards they had a freer passage, the price was very reasonable all the rest of that year.

The public fires which were made on these occasions, as I have calculated it, must necessarily have cost the city about 200 chalders of coals a week, if they had continued, which was indeed a very great quantity; but as it was thought necessary, nothing was spared. However, as some of the physicians cried them down, they were not kept alight above four or five days. The fires were ordered thus:—

One at the Custom House, one at Billingsgate, one at Queenhith, and one at the Three Cranes; one in Blackfriars, and one at the gate of Bridewell; one at the corner of Leadenhal Street and Gracechurch; one at the north and one at the south gate of the Royal Exchange; one at Guild Hall, and one at Blackwell Hall gate; one at the Lord Mayor's door in St Helen's, one at the west entrance into St Paul's, and one at the entrance into Bow Church. I do not remember whether there was any at the city gates, but one at the Bridge-foot there was, just by St Magnus Church.

I know some have quarrelled since that at the experiment, and said that there died the more people because of those fires; but I am persuaded those that say so offer no evidence to prove it, neither can I believe it on any account whatever.

Part 4

It remains to give some account of the state of trade at home in England during this dreadful time, and particularly as it relates to the manufactures and the trade in the city. At the first breaking out of the infection there was, as it is easy to suppose, a very great fright among the people, and consequently a general stop of trade, except in provisions and necessaries of life; and even in those things, as there was a vast number of people fled and a very great number always sick, besides the number which died, so there could not be above two-thirds, if above one-half, of the consumption of provisions in the city as used to be.

It pleased God to send a very plentiful year of corn and fruit, but not of hay or grass—by which means bread was cheap, by reason of the plenty of corn. Flesh was cheap, by reason of the scarcity of grass; but butter and cheese were dear for the same reason, and hay in the market just beyond Whitechappel Bars was sold at 4 pound per load. But that affected not the poor. There was a most excessive plenty of all sorts of fruit, such as apples, pears, plums, cherries, grapes, and they were the cheaper because of the want of people; but this made the poor eat them to excess, and this brought them into fluxes, griping of the guts, surfeits, and the like, which often precipitated them into the plague.

But to come to matters of trade. First, foreign exportation being stopped or at least very much interrupted and rendered difficult, a general stop of all those manufactures followed of course which were usually brought for exportation; and though sometimes merchants abroad were importunate for goods, yet little was sent, the passages being so generally stopped that the English ships would not be admitted, as is said already, into their port.

This put a stop to the manufactures that were for exportation in most parts of England, except in some out-ports; and even that was soon stopped, for they all had the plague in their turn. But though this was felt all over England, yet, what was still worse, all intercourse of trade for home consumption of manufactures, especially those which usually circulated through the Londoner's hands, was stopped at once, the trade of the city being stopped.

All kinds of handicrafts in the city, &c., tradesmen and mechanics, were, as I have said before, out of employ; and this occasioned the putting-off and dismissing an innumerable number of journeymen and workmen of all sorts, seeing nothing was done relating to such trades but what might be said to be absolutely necessary.

This caused the multitude of single people in London to be unprovided for, as also families whose living depended upon the labour of the heads of those families; I say, this reduced them to extreme misery; and I must confess it is for the honour of the city of London, and will be for many ages, as long as this is to be spoken of, that they were able to supply with charitable provision the wants of so many thousands of those as afterwards fell sick and were distressed: so that it may be safely averred that nobody perished for want, at least that the magistrates had any notice given them of.

This stagnation of our manufacturing trade in the country would have put the people there to much greater difficulties, but that the master-workmen, clothiers and others, to the uttermost of their stocks and strength, kept on making their goods to keep the poor at work, believing that soon as the sickness should abate they would have a quick demand in proportion to the decay of their trade at that time. But as none but those masters that were rich could do thus, and that many were poor and not able, the manufacturing trade in England suffered greatly, and the poor were pinched all over England by the calamity of the city of London only.

It is true that the next year made them full amends by another terrible calamity upon the city; so that the city by one calamity impoverished and weakened the country, and by another calamity, even terrible too of its kind, enriched the country and made them again amends; for an infinite quantity of household Stuff, wearing apparel, and other things, besides whole warehouses filled with merchandise and manufactures such as come from all parts of England, were consumed in the fire of London the next year after this terrible visitation. It is incredible what a trade this made all over the whole kingdom, to make good the want and to supply that loss; so that, in short, all the manufacturing hands in the nation were set on work, and were little enough for several years to supply the market and answer the demands. All foreign markets also were empty of our goods by the stop which had been occasioned by the plague, and before an open trade was allowed again; and the prodigious demand at home falling in, joined to make a quick vent for all sort of goods; so that there never was known such a trade all over England for the time as was in the first seven years after the plague, and after the fire of London.

Part 5

It remains now that I should say something of the merciful part of this terrible judgement. The last week in September, the plague being come to its crisis, its fury began to assuage. I remember my friend Dr Heath, coming to see me the week before, told me he was sure that the violence of it would assuage in a few days; but when I saw the weekly bill of that week, which was the highest of the whole year, being 8297 of all diseases, I upbraided him with it, and asked him what he had made his judgement from. His answer, however, was not so much to seek as I thought it would have been. 'Look you,' says he, 'by the number which are at this time sick and infected, there should have been twenty thousand dead the last week instead of eight thousand, if the inveterate mortal contagion had been as it was two weeks ago; for then it ordinarily killed in two or three days, now not under eight or ten; and then not above one in five recovered, whereas I have observed that now not above two in five miscarry. And, observe it from me, the next bill will decrease, and you will see many more people recover than used to do; for though a vast multitude are now everywhere infected, and as many every day fall sick, yet there will not so many die as there did, for the malignity of the distemper is abated';—adding that he began now to hope, nay, more than hope, that the infection had passed its crisis and was going off; and accordingly so it was, for the next week being, as I said, the last in September, the bill decreased almost two thousand.

It is true the plague was still at a frightful height, and the next bill was no less than 6460, and the next to that, 5720; but still my friend's observation was just, and it did appear the people did recover faster and more in number than they used to do; and indeed, if it had not been so, what had been the condition of the city of London? For, according to my friend, there were not fewer than 60,000 people at that time infected, whereof, as above, 20,477 died, and near 40,000 recovered; whereas, had it been as it was before, 50,000 of that number would very probably have died, if not more, and 50,000 more would have sickened; for, in a word, the whole mass of people began to sicken, and it looked as if none would escape.

But this remark of my friend's appeared more evident in a few weeks more, for the decrease went on, and another week in October it decreased 1843, so that the number dead of the plague was but 2665; and the next week it decreased 1413 more, and yet it was seen plainly that there was abundance of people sick, nay, abundance more than ordinary, and abundance fell sick every day but (as above) the malignity of the disease abated.

Such is the precipitant disposition of our people (whether it is so or not all over the world, that's none of my particular business to inquire), but I saw it apparently here, that as upon the first fright of the infection they shunned one another, and fled from one another's houses and from the city with an unaccountable and, as I thought, unnecessary fright, so now, upon this notion spreading, viz., that the distemper was not so catching as formerly, and that if it was catched it was not so mortal, and seeing abundance of people who really fell sick recover again daily, they took to such a precipitant courage, and grew so entirely regardless of themselves and of the infection, that they made no more of the plague than of an ordinary fever, nor indeed so much. They not only went boldly into company with those who had tumours and carbuncles upon them that were running, and consequently contagious, but ate and drank with them, nay, into their houses to visit them, and even, as I was told, into their very chambers where they lay sick.

This I could not see rational. My friend Dr Heath allowed, and it was plain to experience, that the distemper was as catching as ever, and as many fell sick, but only he alleged that so many of those that fell sick did not die; but I think that while many did die, and that at best the distemper itself was very terrible, the sores and swellings very tormenting, and the danger of death not left out of the circumstances of sickness, though not so frequent as before; all those things, together with the exceeding tediousness of the cure, the loathsomeness of the disease, and many other articles, were enough to deter any man living from a dangerous mixture with the sick people, and make them as anxious almost to avoid the infections as before.

Nay, there was another thing which made the mere catching of the distemper frightful, and that was the terrible burning of the caustics which the surgeons laid on the swellings to bring them to break and to run, without which the danger of death was very great, even to the last. Also, the insufferable torment of the swellings, which, though it might not make people raving and distracted, as they were before, and as I have given several instances of already, yet they put the patient to inexpressible torment; and those that fell into it, though they did escape with life, yet they made bitter complaints of those that had told them there was no danger, and sadly repented their rashness and folly in venturing to run into the reach of it.

Nor did this unwary conduct of the people end here, for a great many that thus cast off their cautions suffered more deeply still, and though many escaped, yet many died; and at least it had this public mischief attending it, that it made the decrease of burials slower than it would otherwise have been. For as this notion ran like lightning through the city, and people's heads were possessed with it, even as soon as the first great decrease in the bills appeared, we found that the two next bills did not decrease in proportion; the reason I take to be the people's running so rashly into danger, giving up all their former cautions and care, and all the shyness which they used to practise, depending that the sickness would not reach them—or that if it did, they should not die.

The physicians opposed this thoughtless humour of the people with all their might, and gave out printed directions, spreading them all over the city and suburbs, advising the people to continue reserved, and to use still the utmost caution in their ordinary conduct, notwithstanding the decrease of the distemper, terrifying them with the danger of bringing a relapse upon the whole city, and telling them how such a relapse might be more fatal and dangerous than the whole visitation that had been already; with many arguments and reasons to explain and prove that part to them, and which are too long to repeat here.

But it was all to no purpose; the audacious creatures were so possessed with the first joy and so surprised with the satisfaction of seeing a vast decrease in the weekly bills, that they were impenetrable by any new terrors, and would not be persuaded but that the bitterness of death was past; and it was to no more purpose to talk to them than to an east wind; but they opened shops, went about streets, did business, and conversed with anybody that came in their way to converse with, whether with business or without, neither inquiring of their health or so much as being apprehensive of any danger from them, though they knew them not to be sound.

This imprudent, rash conduct cost a great many their lives who had with great care and caution shut themselves up and kept retired, as it were, from all mankind, and had by that means, under God's providence, been preserved through all the heat of that infection.

This rash and foolish conduct, I say, of the people went so far that the ministers took notice to them of it at last, and laid before them both the folly and danger of it; and this checked it a little, so that they grew more cautious. But it had another effect, which they could not check; for as the first rumour had spread not over the city only, but into the country, it had the like effect: and the people were so tired with being so long from London, and so eager to come back, that they flocked to town without fear or forecast, and began to show themselves in the streets as if all the danger was over. It was indeed surprising to see it, for though there died still from 1000 to 1800 a week, yet the people flocked to town as if all had been well.

The consequence of this was, that the bills increased again 400 the very first week in November; and if I might believe the physicians, there was above 3000 fell sick that week, most of them new-comers, too.

One John Cock, a barber in St Martin's-le-Grand, was an eminent example of this; I mean of the hasty return of the people when the plague was abated. This John Cock had left the town with his whole family, and locked up his house, and was gone in the country, as many others did; and finding the plague so decreased in November that there died but 905 per week of all diseases, he ventured home again. He had in his family ten persons; that is to say, himself and wife, five children, two apprentices, and a maidservant. He had not returned to his house above a week, and began to open his shop and carry on his trade, but the distemper broke out in his family, and within about five days they all died, except one; that is to say, himself. his wife, all his five children, and his two apprentices; and only the maid remained alive.

But the mercy of God was greater to the rest than we had reason to expect; for the malignity (as I have said) of the distemper was spent, the contagion was exhausted, and also the winter weather came on apace, and the air was clear and cold, with sharp frosts; and this increasing still, most of those that had fallen sick recovered, and the health of the city began to return. There were indeed some returns of the distemper even in the month of December, and the bills increased near a hundred; but it went off again, and so in a short while things began to return to their own channel. And wonderful it was to see how populous the city was again all on a sudden, so that a stranger could not miss the numbers that were lost. Neither was there any miss of the inhabitants as to their dwellings—few or no empty houses were to be seen, or if there were some, there was no want of tenants for them.

I wish I could say that as the city had a new face, so the manners of the people had a new appearance. I doubt not but there were many that retained a sincere sense of their deliverance, and were that heartily thankful to that Sovereign Hand that had protected them in so dangerous a time; it would be very uncharitable to judge otherwise in a city so populous, and where the people were so devout as they were here in the time of the visitation itself; but except what of this was to be found in particular families and faces, it must be acknowledged that the general practice of the people was just as it was before, and very little difference was to be seen.

Some, indeed, said things were worse; that the morals of the people declined from this very time; that the people, hardened by the danger they had been in, like seamen after a storm is over, were more wicked and more stupid, more bold and hardened, in their vices and immoralities than they were before; but I will not carry it so far neither. It would take up a history of no small length to give a particular of all the gradations by which the course of things in this city came to be restored again, and to run in their own channel as they did before.

Some parts of England were now infected as violently as London had been; the cities of Norwich, Peterborough, Lincoln, Colchester, and other places were now visited; and the magistrates of London began to set rules for our conduct as to corresponding with those cities. It is true we could not pretend to forbid their people coming to London, because it was impossible to know them asunder; so, after many consultations, the Lord Mayor and Court of Aldermen were obliged to drop it. All they could do was to warn and caution the people not to entertain in their houses or converse with any people who they knew came from such infected places.

But they might as well have talked to the air, for the people of London thought themselves so plague-free now that they were past all admonitions; they seemed to depend upon it that the air was restored, and that the air

was like a man that had had the smallpox, not capable of being infected again. This revived that notion that the infection was all in the air, that there was no such thing as contagion from the sick people to the sound; and so strongly did this whimsy prevail among people that they ran all together promiscuously, sick and well. Not the Mahometans, who, prepossessed with the principle of predestination, value nothing of contagion, let it be in what it will, could be more obstinate than the people of London; they that were perfectly sound, and came out of the wholesome air, as we call it, into the city, made nothing of going into the same houses and chambers, nay, even into the same beds, with those that had the distemper upon them, and were not recovered.

Some, indeed, paid for their audacious boldness with the price of their lives; an infinite number fell sick, and the physicians had more work than ever, only with this difference, that more of their patients recovered; that is to say, they generally recovered, but certainly there were more people infected and fell sick now, when there did not die above a thousand or twelve hundred in a week, than there was when there died five or six thousand a week, so entirely negligent were the people at that time in the great and dangerous case of health and infection, and so ill were they able to take or accept of the advice of those who cautioned them for their good.

The people being thus returned, as it were, in general, it was very strange to find that in their inquiring after their friends, some whole families were so entirely swept away that there was no remembrance of them left, neither was anybody to be found to possess or show any title to that little they had left; for in such cases what was to be found was generally embezzled and purloined, some gone one way, some another.

It was said such abandoned effects came to the king, as the universal heir; upon which we are told, and I suppose it was in part true, that the king granted all such, as deodands, to the Lord Mayor and Court of Aldermen of London, to be applied to the use of the poor, of whom there were very many. For it is to be observed, that though the occasions of relief and the objects of distress were very many more in the time of the violence of the plague than now after all was over, yet the distress of the poor was more now a great deal than it was then, because all the sluices of general charity were now shut. People supposed the main occasion to be over, and so stopped their hands; whereas particular objects were still very moving, and the distress of those that were poor was very great indeed.

Though the health of the city was now very much restored, yet foreign trade did not begin to stir, neither would foreigners admit our ships into their ports for a great while. As for the Dutch, the misunderstandings between our court and them had broken out into a war the year before, so that our trade that way was wholly interrupted; but Spain and Portugal, Italy and Barbary, as also Hamburg and all the ports in the Baltic, these were all shy of us a great while, and would not restore trade with us for many months.

The distemper sweeping away such multitudes, as I have observed, many if not all the out-parishes were obliged to make new burying-grounds, besides that I have mentioned in Bunhill Fields, some of which were continued, and remain in use to this day. But others were left off, and (which I confess I mention with some reflection) being converted into other uses or built upon afterwards, the dead bodies were disturbed, abused, dug up again, some even before the flesh of them was perished from the bones, and removed like dung or rubbish to other places. Some of those which came within the reach of my observation are as follow:

(1) A piece of ground beyond Goswell Street, near Mount Mill, being some of the remains of the old lines or fortifications of the city, where abundance were buried promiscuously from the parishes of Aldersgate, Clerkenwell, and even out of the city. This ground, as I take it, was since made a physic garden, and after that has been built upon.

(2) A piece of ground just over the Black Ditch, as it was then called, at the end of Holloway Lane, in Shoreditch parish. It has been since made a yard for keeping hogs, and for other ordinary uses, but is quite out of use as a burying-ground.

(3) The upper end of Hand Alley, in Bishopsgate Street, which was then a green field, and was taken in particularly for Bishopsgate parish, though many of the carts out of the city brought their dead thither also, particularly out of the parish of St All-hallows on the Wall. This place I cannot mention without much regret. It was, as I remember, about two or three years after the plague was ceased that Sir Robert Clayton came to be possessed of the ground. It was reported, how true I know not, that it fell to the king for want of heirs, all those who had any right to it being carried off by the pestilence, and that Sir Robert Clayton obtained a grant of it from King Charles II. But however he came by it, certain it is the ground was let out to build on, or built upon, by his order. The first house built upon it was a large fair house, still standing, which faces the street or way now called Hand Alley which, though called an alley, is as wide as a street. The houses in the same row with that house northward are built on the very same ground where the poor people were buried, and the bodies, on opening the ground for the foundations, were dug

up, some of them remaining so plain to be seen that the women's skulls were distinguished by their long hair, and of others the flesh was not quite perished; so that the people began to exclaim loudly against it, and some suggested that it might endanger a return of the contagion; after which the bones and bodies, as fast as they came at them, were carried to another part of the same ground and thrown all together into a deep pit, dug on purpose, which now is to be known in that it is not built on, but is a passage to another house at the upper end of Rose Alley, just against the door of a meeting-house which has been built there many years since; and the ground is palisadoed off from the rest of the passage, in a little square; there lie the bones and remains of near two thousand bodies, carried by the dead carts to their grave in that one year.

(4) Besides this, there was a piece of ground in Moorfields; by the going into the street which is now called Old Bethlem, which was enlarged much, though not wholly taken in on the same occasion.

[N.B.—The author of this journal lies buried in that very ground, being at his own desire, his sister having been buried there a few years before.]

(5) Stepney parish, extending itself from the east part of London to the north, even to the very edge of Shoreditch Churchyard, had a piece of ground taken in to bury their dead close to the said churchyard, and which for that very reason was left open, and is since, I suppose, taken into the same churchyard. And they had also two other burying-places in Spittlefields, one where since a chapel or tabernacle has been built for ease to this great parish, and another in Petticoat Lane.

There were no less than five other grounds made use of for the parish of Stepney at that time: one where now stands the parish church of St Paul, Shadwell, and the other where now stands the parish church of St John's at Wapping, both which had not the names of parishes at that time, but were belonging to Stepney parish.

I could name many more, but these coming within my particular knowledge, the circumstance, I thought, made it of use to record them. From the whole, it may be observed that they were obliged in this time of distress to take in new burying-grounds in most of the out-parishes for laying the prodigious numbers of people which died in so short a space of time; but why care was not taken to keep those places separate from ordinary uses, that so the bodies might rest undisturbed, that I cannot answer for, and must confess I think it was wrong. Who were to blame I know not.

I should have mentioned that the Quakers had at that time also a burying-ground set apart to their use, and which they still make use of; and they had also a particular dead-cart to fetch their dead from their houses; and the famous Solomon Eagle, who, as I mentioned before, had predicted the plague as a judgement, and ran naked through the streets, telling the people that it was come upon them to punish them for their sins, had his own wife died the very next day of the plague, and was carried, one of the first in the Quakers' dead-cart, to their new burying-ground.

I might have thronged this account with many more remarkable things which occurred in the time of the infection, and particularly what passed between the Lord Mayor and the Court, which was then at Oxford, and what directions were from time to time received from the Government for their conduct on this critical occasion. But really the Court concerned themselves so little, and that little they did was of so small import, that I do not see it of much moment to mention any part of it here: except that of appointing a monthly fast in the city and the sending the royal charity to the relief of the poor, both which I have mentioned before.

Great was the reproach thrown on those physicians who left their patients during the sickness, and now they came to town again nobody cared to employ them. They were called deserters, and frequently bills were set up upon their doors and written, 'Here is a doctor to be let', so that several of those physicians were fain for a while to sit still and look about them, or at least remove their dwellings, and set up in new places and among new acquaintance. The like was the case with the clergy, whom the people were indeed very abusive to, writing verses and scandalous reflections upon them, setting upon the church-door, 'Here is a pulpit to be let', or sometimes, 'to be sold', which was worse.

It was not the least of our misfortunes that with our infection, when it ceased, there did not cease the spirit of strife and contention, slander and reproach, which was really the great troubler of the nation's peace before. It was said to be the remains of the old animosities, which had so lately involved us all in blood and disorder. But as the late Act of Indemnity had laid asleep the quarrel itself, so the Government had recommended family and personal peace upon all occasions to the whole nation.

But it could not be obtained; and particularly after the ceasing of the plague in London, when any one that had seen the condition which the people had been in, and how they caressed one another at that time, promised to have more charity for the future, and to raise no more reproaches; I say, any one that had seen them then would have thought they would have come together with another spirit at last. But, I say, it could not be obtained. The quarrel remained; the Church and the Presbyterians were incompatible. As soon as the plague was removed, the Dissenting ousted ministers who had supplied the pulpits which were deserted by the incumbents retired; they could expect no other but that they should immediately fall upon them and harass them with their penal laws, accept their preaching while they were sick, and persecute them as soon as they were recovered again; this even we that were of the Church thought was very hard, and could by no means approve of it.

But it was the Government, and we could say nothing to hinder it; we could only say it was not our doing, and we could not answer for it.

On the other hand, the Dissenters reproaching those ministers of the Church with going away and deserting their charge, abandoning the people in their danger, and when they had most need of comfort, and the like: this we could by no means approve, for all men have not the same faith and the same courage, and the Scripture commands us to judge the most favourably and according to charity.

A plague is a formidable enemy, and is armed with terrors that every man is not sufficiently fortified to resist or prepared to stand the shock against. It is very certain that a great many of the clergy who were in circumstances to do it withdrew and fled for the safety of their lives; but 'tis true also that a great many of them stayed, and many of them fell in the calamity and in the discharge of their duty.

It is true some of the Dissenting turned-out ministers stayed, and their courage is to be commended and highly valued—but these were not abundance; it cannot be said that they all stayed, and that none retired into the country, any more than it can be said of the Church clergy that they all went away. Neither did all those that went away go without substituting curates and others in their places, to do the offices needful and to visit the sick, as far as it was practicable; so that, upon the whole, an allowance of charity might have been made on both sides, and we should have considered that such a time as this of 1665 is not to be paralleled in history, and that it is not the stoutest courage that will always support men in such cases. I had not said this, but had rather chosen to record the courage and religious zeal of those of both sides, who did hazard themselves for the service of the poor people in their distress, without remembering that any failed in their duty on either side. But the want of temper among us has made the contrary to this necessary: some that stayed not only boasting too much of themselves, but reviling those that fled, branding them with cowardice, deserting their flocks, and acting the part of the hireling, and the like. I recommend it to the charity of all good people to look back and reflect duly upon the terrors of the time, and whoever does so well see that it is not an ordinary strength that could support it. It was not like appearing in the head of an army or charging a body of horse in the field, but it was charging Death itself on his pale horse; to stay was indeed to die, and it could be esteemed nothing less, especially as things appeared at the latter end of August and the beginning of September, and as there was reason to expect them at that time; for no man expected, and I dare say believed, that the distemper would take so sudden a turn as it did, and fall immediately two thousand in a week, when there was such a prodigious number of people sick at that time as it was known there was; and then it was that many shifted away that had stayed most of the time before.

Besides, if God gave strength to some more than to others, was it to boast of their ability to abide the stroke, and upbraid those that had not the same gift and support, or ought not they rather to have been humble and thankful if they were rendered more useful than their brethren?

I think it ought to be recorded to the honour of such men, as well clergy as physicians, surgeons, apothecaries, magistrates, and officers of every kind, as also all useful people who ventured their hves in discharge of their duty, as most certainly all such as stayed did to the last degree; and several of all these kinds did not only venture but lose their lives on that sad occasion.

I was once making a list of all such, I mean of all those professions and employments who thus died, as I call it, in the way of their duty; but it was impossible for a private man to come at a certainty in the particulars. I only remember that there died sixteen clergymen, two aldermen, five physicians, thirteen surgeons, within the city and liberties before the beginning of September. But this being, as I said before, the great crisis and extremity of the infection, it can be no complete list. As to inferior people, I think there died six-and-forty constables and head-boroughs in the two parishes of Stepney and Whitechappel; but I could not carry my list oil, for when the violent rage of the distemper in September came upon us, it drove us out of all measures. Men did then no more (lie by tale and by number. They might put out a weekly bill, and call them seven or eight thousand, or what they pleased; 'tis certain they died by heaps, and were buried by heaps, that is to say, without account. And if I might believe some people, who were more abroad and more conversant with those things than I though I was public enough for one that had no more business to do than I had—I say, if I may believe them, there was not many less buried those first three weeks in September than 20,000 per week. However, the others aver the truth of it; yet I rather choose to keep to the public account; seven and eight thousand per week is enough to make good all that I have said of the terror of those times; and it is much to the satisfaction of me that write, as well as those that read, to be able to say that everything is set down with moderation, and rather within compass than beyond it.

Upon all these accounts, I say, I could wish, when we were recovered, our conduct had been more distinguished for charity and kindness in remembrance of the past calamity, and not so much a valuing ourselves upon our boldness in staying, as if all men were cowards that fly from the hand of God, or that those who stay do not sometimes owe their courage to their ignorance, and despising the hand of their Maker—which is a criminal kind of desperation, and not a true courage.

I cannot but leave it upon record that the civil officers, such as constables, head-boroughs, Lord Mayor's and sheriffs'-men, as also parish officers, whose business it was to take charge of the poor, did their duties in general with as much courage as any, and perhaps with more, because their work was attended with more hazards, and lay more among the poor, who were more subject to be infected, and in the most pitiful plight when they were taken with the infection. But then it must be added, too, that a great number of them died; indeed it was scarce possible it should be otherwise.

I have not said one word here about the physic or preparations that we ordinarily made use of on this terrible occasion—I mean we that went frequently abroad and up down street, as I did; much of this was talked of in the books and bills of our quack doctors, of whom I have said enough already. It may, however, be added, that the College of Physicians were daily publishing several preparations, which they had considered of in the process of their practice, and which, being to be had in print, I avoid repeating them for that reason.

One thing I could not help observing: what befell one of the quacks, who published that he had a most excellent preservative against the plague, which whoever kept about them should never be infected or liable to infection. This man, who, we may reasonably suppose, did not go abroad without some of this excellent preservative in his pocket, yet was taken by the distemper, and carried off in two or three days.

I am not of the number of the physic-haters or physic-despisers; on the contrary, I have often mentioned the regard I had to the dictates of my particular friend Dr Heath; but yet I must acknowledge I made use of little or nothing—except, as I have observed, to keep a preparation of strong scent to have ready, in case I met with anything of offensive smells or went too near any burying-place or dead body.

Neither did I do what I know some did: keep the spirits always high and hot with cordials and wine and such things; and which, as I observed, one learned physician used himself so much to as that he could not leave them off when the infection was quite gone, and so became a sot for all his life after.

I remember my friend the doctor used to say that there was a certain set of drugs and preparations which were all certainly good and useful in the case of an infection; out of which, or with which, physicians might make an infinite variety of medicines, as the ringers of bells make several hundred different rounds of music by the changing and order or sound but in six bells, and that all these preparations shall be really very good: Therefore,' said he, 'I do not wonder that so vast a throng of medicines is offered in the present calamity, and almost every physician prescribes or prepares a different thing, as his judgement or experience guides him; but', says my friend, 'let all the prescriptions of all the physicians in London be examined, and it will be found that they are all compounded of the same things, with such variations only as the particular fancy of the doctor leads him to; so that', says he, 'every man, judging a little of his own constitution and manner of his living, and circumstances of his being infected, may direct his own medicines out of the ordinary drugs and preparations. Only that', says he, 'some recommend one thing as most sovereign, and some another. Some', says he, 'think that pill, ruff., which is called itself the antipestilential pill is the best preparation that can be made; others think that Venice treacle is sufficient of itself to resist the contagion; and I', says he, 'think as both these think, viz., that the last is good to take beforehand to prevent it, and the first, if touched, to expel it,' According to this opinion, I several times took Venice treacle, and a sound sweat upon it, and thought myself as well fortified against the infection as any one could be fortified by the power of physic.

As for quackery and mountebanks, of which the town was so full, I listened to none of them, and have observed often since, with some wonder, that for two years after the plague I scarcely saw or heard of one of them about town. Some fancied they were all swept away in the infection to a man, and were for calling it a particular mark of God's vengeance upon them for leading the poor people into the pit of destruction, merely for the lucre of a little money they got by them; but I cannot go that length neither. That abundance of them died is certain—many of them came within the reach of my own knowledge—but that all of them were swept off I much question. I believe rather they fled into the country and tried their practices upon the people there, who were in apprehension of the infection before it came among them.

This, however, is certain, not a man of them appeared for a great while in or about London. There were, indeed, several doctors who published bills recommending their several physical preparations for cleansing the body, as they call it, after the plague, and needful, as they said, for such people to take who had been visited and had been cured; whereas I must own I believe that it was the opinion of the most eminent physicians at that time that the plague was itself a sufficient purge, and that those who escaped the infection needed no physic to cleanse their bodies of any other things; the running sores, the tumours, &c., which were broke and kept open by the directions of the physicians, having sufficiently cleansed them; and that all other distempers, and causes of distempers, were effectually carried off that way; and as the physicians gave this as their opinions wherever they came, the quacks got little business.

There were, indeed, several little hurries which happened after the decrease of the plague, and which, whether they were contrived to fright and disorder the people, as some imagined, I cannot say, but sometimes we were told the plague would return by such a time; and the famous Solomon Eagle, the naked Quaker I have mentioned, prophesied evil tidings every day; and several others telling us that London had not been sufficiently scourged, and that sorer and severer strokes were yet behind. Had they stopped there, or had they descended to particulars, and told us that the city should the next year be destroyed by fire, then, indeed, when we had seen it come to pass, we should not have been to blame to have paid more than a common respect to their prophetic spirits; at least we should have wondered at them, and have been more serious in our inquiries after the meaning of it, and whence they had the foreknowledge. But as they generally told us of a relapse into the plague, we have had no concern since that about them; yet by those frequent clamours, we were all kept with some kind of apprehensions constantly upon us; and if any died suddenly, or if the spotted fevers at any time increased, we were presently alarmed; much more if the number of the plague increased, for to the end of the year there were always between 200 and 300 of the plague. On any of these occasions, I say, we were alarmed anew.

Those who remember the city of London before the fire must remember that there was then no such place as we now call Newgate Market, but that in the middle of the street which is now called Blow-bladder Street, and which had its name from the butchers, who used to kill and dress their sheep there (and who, it seems, had a custom to blow up their meat with pipes to make it look thicker and fatter than it was, and were punished there for it by the Lord Mayor); I say, from the end of the street towards Newgate there stood two long rows of shambles for the selling meat.

It was in those shambles that two persons falling down dead, as they were buying meat, gave rise to a rumour that the meat was all infected; which, though it might affright the people, and spoiled the market for two or three days, yet it appeared plainly afterwards that there was nothing of truth in the suggestion. But nobody can account for the possession of fear when it takes hold of the mind.

However, it Pleased God, by the continuing of the winter weather, so to restore the health of the city that by February following we reckoned the distemper quite ceased, and then we were not so easily frighted again.

There was still a question among the learned, and at first perplexed the people a little: and that was in what manner to purge the house and goods where the plague had been, and how to render them habitable again, which had been left empty during the time of the plague. Abundance of perfumes and preparations were prescribed by physicians, some of one kind and some of another, in which the people who listened to them put themselves to a great, and indeed, in my opinion, to an unnecessary expense; and the poorer people, who only set open their windows night and day, burned brimstone, pitch, and gunpowder, and such things in their rooms, did as well as the best; nay, the eager people who, as I said above, came home in haste and at all hazards, found little or no inconvenience in their houses, nor in the goods, and did little or nothing to them.

However, in general, prudent, cautious people did enter into some measures for airing and sweetening their houses, and burned perfumes, incense, benjamin, rozin, and sulphur in their rooms close shut up, and then let the air carry it all out with a blast of gunpowder; others caused large fires to be made all day and all night for several days and nights; by the same token that two or three were pleased to set their houses on fire, and so effectually sweetened them by burning them down to the ground; as particularly one at Ratcliff, one in Holbourn, and one at Westminster; besides two or three that were set on fire, but the fire was happily got out again before it went far enough to bum down the houses; and one citizen's servant, I think it was in Thames Street, carried so much gunpowder into his master's house, for clearing it of the infection, and managed it so foolishly, that he blew up part of the roof of the house. But the time was not fully come that the city was to he purged by fire, nor was it far off; for within nine months more I saw it all lying in ashes; when, as some of our quacking philosophers pretend, the seeds of the plague were entirely destroyed, and not before; a notion too ridiculous to speak of here; since, had the seeds of the plague remained in the houses, not to be destroyed but by fire, how has it been that they have not since broken out, seeing all those buildings in the suburbs and liberties, all in the great parishes of Stepney, Whitechappel, Aldgate, Bishopsgate, Shoreditch, Cripplegate, and St Giles, where the fire never came, and where the plague raged with the greatest violence, remain still in the same condition they were in before?

But to leave these things just as I found them, it was certain that those people who were more than ordinarily cautious of their health, did take particular directions for what they called seasoning of their houses, and abundance of costly things were consumed on that account which I cannot but say not only seasoned those houses, as they desired, but filled the air with very grateful and wholesome smells which others had the share of the benefit of as well as those who were at the expenses of them.

And yet after all, though the poor came to town very precipitantly, as I have said, yet I must say the rich made no such haste. The men of business, indeed, came up, but many of them did not bring their families to town till the spring came on, and that they saw reason to depend upon it that the plague would not return.

The Court, indeed, came up soon after Christmas, but the nobility and gentry, except such as depended upon and had employment under the administration, did not come so soon.

I should have taken notice here that, notwithstanding the violence of the plague in London and in other places, yet it was very observable that it was never on board the fleet; and yet for some time there was a strange press in the river, and even in the streets, for seamen to man the fleet. But it was in the beginning of the year, when the plague was scarce begun, and not at all come down to that part of the city where they usually press for seamen; and though a war with the Dutch was not at all grateful to the people at that time, and the seamen went with a kind of reluctancy into the service, and many complained of being dragged into it by force, yet it proved in the event a happy violence to several of them, who had probably perished in the general calamity, and who, after the summer service was over, though they had cause to lament the desolation of their families—who, when they came back, were many of them in their graves—yet they had room to be thankful that they were carried out of the reach of it, though so much against their wills. We indeed had a hot war with the Dutch that year, and one very great engagement at sea in which the Dutch were worsted, but we lost a great many men and some ships. But, as I observed, the plague was not in the fleet, and when they came to lay up the ships in the river the violent part of it began to abate.

I would be glad if I could close the account of this melancholy year with some particular examples historically; I mean of the thankfulness to God, our preserver, for our being delivered from this dreadful calamity. Certainly the circumstance of the deliverance, as well as the terrible enemy we were delivered from, called upon the whole nation for it. The circumstances of the deliverance were indeed very remarkable, as I have in part mentioned already, and particularly the dreadful condition which we were all in when we were to the surprise of the whole town made joyful with the hope of a stop of the infection.

Nothing but the immediate finger of God, nothing but omnipotent power, could have done it. The contagion despised all medicine; death raged in every corner; and had it gone on as it did then, a few weeks more would have cleared the town of all, and everything that had a soul. Men everywhere began to despair; every heart failed them for fear; people were made desperate through the anguish of their souls, and the terrors of death sat in the very faces and countenances of the people.

In that very moment when we might very well say, 'Vain was the help of man'—I say, in that very moment it pleased God, with a most agreeable surprise, to cause the fury of it to abate, even of itself; and the malignity declining, as I have said, though infinite numbers were sick, yet fewer died, and the very first weeks' bill decreased 1843; a vast number indeed!

It is impossible to express the change that appeared in the very countenances of the people that Thursday morning when the weekly bill came out. It might have been perceived in their countenances that a secret surprise and smile of joy sat on everybody's face. They shook one another by the hands in the streets, who would hardly go on the same side of the way with one another before. Where the streets were not too broad they would open their windows and call from one house to another, and ask how they did, and if they had heard the good news that the plague was abated. Some would return, when they said good news, and ask, 'What good news?' and when they answered that the plague was abated and the bills decreased almost two thousand, they would cry out, 'God be praised I' and would weep aloud for joy, telling them they had heard nothing of it; and such was the joy of the people that it was, as it were, life to them from the grave. I could almost set down as many extravagant things done in the excess of their joy as of their grief; but that would be to lessen the value of it.

I must confess myself to have been very much dejected just before this happened; for the prodigious number that were taken sick the week or two before, besides those that died, was such, and the lamentations were so great everywhere, that a man must have seemed to have acted even against his reason if he had so much as expected to escape; and as there was hardly a house but mine in all my neighbourhood but was infected, so had it gone on it would not have been long that there would have been any more neighbours to be infected. Indeed it is hardly credible what dreadful havoc the last three weeks had made, for if I might believe the person whose calculations I always found very well grounded, there were not less than 30,000 people dead and near 100.000 fallen sick in the three weeks I speak of; for the number that sickened was surprising, indeed it was astonishing, and those whose courage upheld them all the time before, sank under it now.

In the middle of their distress, when the condition of the city of London was so truly calamitous, just then it pleased God—as it were by His immediate hand to disarm this enemy; the poison was taken out of the sting. It was wonderful; even the physicians themselves were surprised at it. Wherever they visited they found their patients better; either they had sweated kindly, or the tumours were broke, or the carbuncles went down and the inflammations round them changed colour, or the fever was gone, or the violent headache was assuaged, or some good symptom was in the case; so that in a few days everybody was recovering, whole families that were infected and down, that had ministers praying with them, and expected death every hour, were revived and healed, and none died at all out of them.

Nor was this by any new medicine found out, or new method of cure discovered, or by any experience in the operation which the physicians or surgeons attained to; but it was evidently from the secret invisible hand of Him that had at first sent this disease as a judgement upon us; and let the atheistic part of mankind call my saying what they please, it is no enthusiasm; it was acknowledged at that time by all mankind. The disease was enervated and its malignity spent; and let it proceed from whencesoever it will, let the philosophers search for reasons in nature to account for it by, and labour as much as they will to lessen the debt they owe to their Maker, those physicians who had the least share of religion in them were obliged to acknowledge that it was all supernatural, that it was extraordinary, and that no account could be given of it.

If I should say that this is a visible summons to us all to thankfulness, especially we that were under the terror of its increase, perhaps it may be thought by some, after the sense of the thing was over, an officious canting of religious things, preaching a sermon instead of writing a history, making myself a teacher instead of giving my observations of things; and this restrains me very much from going on here as I might otherwise do. But if ten lepers Were healed, and but one returned to give thanks, I desire to be as that one, and to be thankful for myself.

Nor will I deny but there were abundance of people who, to all appearance, were very thankful at that time; for their mouths were stopped, even the mouths of those whose hearts were not extraordinary long affected with it. But the impression was so strong at that time that it could not be resisted; no, not by the worst of the people.

It was a common thing to meet people in the street that were strangers, and that we knew nothing at all of, expressing their surprise. Going one day through Aldgate, and a pretty many people being passing and repassing, there comes a man out of the end of the Minories, and looking a little up the street and down, he throws his hands abroad, 'Lord, what an alteration is here I Why, last week I came along here, and hardly anybody was to he seen.' Another man—I heard him—adds to his words, "Tis all wonderful; 'tis all a dream.' 'Blessed be God,' says a third man, d and let us give thanks to Him, for 'tis all His own doing, human help and human skill was at an end.' These were all strangers to one another. But such salutations as these were frequent in the street every day; and in spite of a loose behaviour, the very common people went along the streets giving God thanks for their deliverance.

It was now, as I said before, the people had cast off all apprehensions, and that too fast; indeed we were no more afraid now to pass by a man with a white cap upon his head, or with a doth wrapt round his neck, or with his leg limping, occasioned by the sores in his groin, all which were frightful to the last degree, but the week before. But now the street was full of them, and these poor recovering creatures, give them their due, appeared very sensible of their unexpected deliverance; and I should wrong them very much if I should not acknowledge that I believe many of them were really thankful. But I must own that, for the generality of the people, it might too justly be said of them as was said of the children of Israel after their being delivered from the host of Pharaoh, when they passed the Red Sea, and looked back and saw the Egyptians overwhelmed in the water: viz., that they sang His praise, but they soon forgot His works.

I can go no farther here. I should be counted censorious, and perhaps unjust, if I should enter into the unpleasing work of reflecting, whatever cause there was for it, upon the unthankfulness and return of all manner of wickedness among us, which I was so much an eye-witness of myself. I shall conclude the account of this calamitous year therefore with a coarse but sincere stanza of my own, which I placed at the end of my ordinary memorandums the same year they were written:—

A dreadful plague in London was
In the year sixty-five.
Which swept an hundred thousand souls
Away; yet I alive!

H. F

* * *

Does the Bubonic Plague still exist, in remote areas? Yes. The WHO (World Health Organization) reports it is commonly found in: Madagascar; the Democratic Republic of the Congo and Peru.

Does the potential for Bubonic Plague still exist worldwide? Yes, in remote areas.

Has it been found in the United States, recently? Yes, on an irregular basis, in the mountain west: west Texas; New Mexico; Colorado and elsewhere in barren, largely unpopulated regions.

The Huffington Post, on the internet, carried this headline and article July 25, 2021: "Plague Found in 6 Colorado Counties After Girl's Death."

"Health officials in Colorado are asking people to take precautions after plague was detected in six counties, including where a 10-year-old girl recently died from causes associated with the infectious disease, which can turn serious if not immediately treated."

The article also said, "In Colorado, we expect to have fleas test positive for plague during the summer months," said Jennifer House, deputy state epidemiologist and public health veterinarian for the CDPHB (Colorado Department of Public Health and Environment).

Officials urged citizens not to kill prairie dogs on their property, as this could increase the risk of plague transmission if it is present. Prairie dogs are very susceptible to plague and are good indicators of the disease if they suddenly disappear.

2.

The Spanish Flu, 1918-1919

WORLD-WIDE:
NO MASS VACCINES, NO ANTIBIOTICS ... NO HOPE.

Before there was AIDS and Ebola, before Covid-19, there was the Spanish flu pandemic of 1918-1919.

In *Pale Rider: The Spanish Flu of 918 and How It Changed the World*, Laura Spinney offers the perfect simile about an epidemic:

> An epidemic, like a forest fire, depends on "fuel"—that is, individuals susceptible to infection. It grows exponentially from a few initial cases—the "spark"—because those cases are surrounded by a vast pool of susceptible individuals. Over time, however, that pool shrinks as people either die or recover and acquire immunity If you were to draw a graph of an epidemic, therefore, with "number of new cases" on the vertical axis, and "time" on the horizontal axis, you'd be looking at normal distribution, or bell curve.

And, she writes ...

> This is the classic form of an epidemic, though countless variations are possible—the curve's height or width may vary,

for example, or it may have more than one peak. The basic form remains recognizable, which means that it can be described in mathematical terms. In the twenty-first century, the mathematical modeling of epidemics is highly sophisticated but scientists had already begun think that way in 1918. Two years earlier, in his "theory of happenings," the British malaria expert and Nobel laureate Ronald Ross had come up with a set of differential equations that could help determine, at any given time, the proportion of a population that was infected, the proportion that was susceptible, and rate of conversion within the two (with some diseases, infected individuals could return to the susceptible group on recovery). A happening, according to Ross's definition, was anything that spread through a population, be it a germ, a rumour or a fashion, (pp. 95-96).

Ross's idea of "more than one peak," is happening currently (August and later, 2021) with the spikes of Covid-19 due to the Delta variant of Covid-19. The Delta variant is markedly more dangerous—a person with the "original" Covid-19 might infect two or three others; a victim of the Delta variation might infect as many as nine others.

The Spanish Flu virus—like the 2020-2021 Covid-19—was airborne—anyone could get infected by standing too close to another person—who might sneeze or cough or simply breathe on an unsuspecting individual.

The difference, however, in 1918-1919: there were no mass vaccinations available, for the millions around the world who would need protection, during those years.

A brief endorsement for a book, to be used on the back jacket of a hardcover edition or on the back cover of a paperback edition and in advertising and promotion of the book, is called a *blurb* in book publishing. They are always positive; a writer or book editor may ask a writer-friend of the author or an expert for a blurb. (Commonly, a conversation between an author and his or her editor might go like this: "who can we get to blurb this?")

There are three blurbs on the back cover of the paperback second edition of *America's Forgotten Pandemic: The influenza of 1918 by* Alfred W. Crosby. The shortest is by Guido Majno of *Natural History* and is worth reprinting. Italics are added.

> (This) is a fine, galloping account of the influenza pandemic that killed some 25 million people in less than a year. In some ways *it was a page from the Middle Ages bound into the twentieth century.* No plague ever killed so many people in so short a time.

> *A page from the middle ages bound into the twentieth century. An estimated one-third of the entire population of the world was affected; ill, terribly ill with the flu or dying or dead.*

The first case of the flu, in America, was not on the east coast, where citizens of the U.S. might be expected to be infected by incoming travelers from Europe, aboard ships, but in the America heartland. In Camp Funston, in Kansas, a major army training base, for troops of the American Expeditionary Forces heading for World War One in Europe. There Albert Gitchell, an Army cook, was the first, March 4, 1918, then a flood of 522 men reported sick with symptoms which became standard: headaches; fever; backaches. Then there was a wave of pneumonia, 233 cases and 48 fatalities.

Camps Oglethorpe, Gordon, Grant, Lewis, Sherman, Doniphan, Fremont, Hancock, Kearney, Logan, McClellan, Sevier, Shelby and others had epidemics of influenza in March and April of that year, but little attention was paid.

Some doctors noticed that when autopsies were performed, the lungs of the dead were full of fluid.

This new illness was killing the young and healthy and the old and presumably infirm in significant numbers.

But why the young and healthy?

The young—servicemen at first—got the illness; fevers, aches and often died in 48 hours.

Why?

Why were their lungs full of fluid?

From those months—May, 1918, and afterwards, the disease spread "from army to army with ease," Alfred Crosby writes.

At the same time, doctors in London analyzed available statistical records: the flu killed only a few of those affected, but nearly half were between 20 and 45.

Some answers are obvious now: army camps, with tents or barracks; marching formations; chow lines and other close-order situations were ideal for spreading influenza.

Overcrowded hospitals and medical camps also contributed to bacterial super-infections.

And, as Crosby writes,

"Spanish Influenza had rounded the globe in four months and following an appearance in the United States and fully earned a promotion from epidemic to pandemic. It has infected so many that, for all its mildness it doubtlessly had killed tens of thousands already. In its next wave it would kill millions."

The first wave of the flu lasted from the first quarter of 1918 and was mild; the first wave of the flu, however, caused a major disruption of operations during World War I, including infecting three quarters of the French troops and half the British forces.

In May, 1918, the flu had hit the German armies, who called it *Blitzkatarrh*.

In *Pandemic 1918: Eyewitness Accounts from the Greatest Medical Holocaust in Modern History*, Catharine Arnold writes;

> By May, influenza had crossed effortlessly over "No Man's Land" to hit the German army; known to Germans as *Blitzkatarrh*, the disease affected 139,000 men during Jume and peaked in early July. Lasting from four to six days on average, it left troops debilitated and brought the German army to a state of near exhaustion. In late J\me, German commander Eric von Ludendorff noted that over 2,000 men in each division were suffering from influenza, that the supply system was breaking down, and the troops were starving. As the German high command struggled to replace more than 900,000 casualties, influenza put ever-increasing numbers of German soldiers out of commission. By late July, Ludendorff was blaming influenza for halting the German attack.
>
> "Our army suffered. Influenza was rampant," von Ludendorff wrote in his memoirs."It was a grievous business having to listen every morning to the chief of staffs recital of the number of influenza cases, and their complaints about the weakness of their troops if the English attacked again." Influenza had brought the all-conquering German army to its knees, while the Allies, strike too, took advantage of their enemy's weakness to re-group.

The name Spanish Influenza is a misnomer; during World War I, news from Germany, and other opposing nations, was embargoed; Spain was neutral and thus news of infections could be distributed internationally from there. But cartoons were published showing "The Spanish Lady" with a death's head face, black flamenco gown, mantilla and fan. (Many believed her to be a prostitute, infecting everyone she met.) The name stuck. The Spanish were offended; they, in turn, blamed their nearest neighbor. In Spain the flu was called The Portuguese Flu, or sometimes the French Flu.

The second wave began in late August, 1918, perhaps originating from Camp Devens, 30 miles from Boston. From there, it spread by troopships and traveled throughout all of North America, Central and South America.

Today we use the term "super spreaders," close-contact events sure to spread a disease—the 1918-1919 Flu, then Covid-19 today. On September 28, 1918 Philadelphia held a "Philadelphia Liberty Loans Parade" to encourage citizens to buy government bonds for World War 1. Subsequently 12,000 died of the flu after attending that event.

During the peak period of September-December, 1918, 292,000 deaths were reported in the United States from the flu.

The second wave was complicated by bacterial pneumonia, which was often the cause of death. This caused the skin to develop two mahogany spots over the cheekbones; the entire face turned blue-black in a few hours, followed by black coloration in the hands. Side effects were loss of hair (experienced by Katherine Anne Porter; her dark hair fell out and later regrew. When it reappeared it was white and remained white throughout the rest of her life). Other side effects were often massive hemorrhages from the nose, bleeding from the ears and delirium.

When doctors saw the mahogany-blue faces and black hands, they knew death was imminent and there was little or nothing they could do.

The approximate timeline of the 1918-1919 Flu is:

June 1914—World War I begins: the assassination of Archduke Franz Ferdinand, heir to the throne of Austria-Hungary, sets off an unstoppable chain of events: World War I, set Germany, Austria-Hungary and Turkey against the Allied powers of Great Britain, France, Russia, China and eventually the United States, which joined the war late.

April 1917—The United States enters the war; German submarines had sunk three U.S. merchant ships in March 1917, with an extensive loss of life. Woodrow Wilson, as President, with the consent of Congress, declared war on Germany and its allies. The United States began a draft of eligible young men and thousands were sent to training campus to prepare for the war in Europe.

March, 1918—Outbreaks of a flu-like Illnesses were reported at Camp Fimston, Kansas and cases grew rapidly. This is generally known as the first outbreak of the flu in America.

September, 1918—The second and more deadly wave appeared at Camp Devens, near Boston. By the end of the month, more than 14,000 soldiers had the flu; there were 757 deaths. One doctor said soldiers "rapidly develop the most vicious type of Pneumonia that had ever been seen. Two hours after admission they have the Mahogany spots over the cheekbones and a few hours later you can begin to see the Cyanosis extending from the ears and spreading all over the face, until it is hard to distinguish the colored man from the white. It is only a matter of a few hours then the death comes and it is simply a struggle for air until they suffocate. It is horrible."

September, 1918—Influenza becomes a reportable disease in New York City and residents were told to stay home or go to a hospital, if necessary

September 29,1918—the Liberty Bond fund drive takes place in Philadelphia, against the advice of doctors.

Thousands line the streets; subsequently 12,000 die of influenza. It was a "super-spreader" event, although that term was not used at that time.

Fall, 1918—The second wave of the flu peaked in the United States. The peak was exacerbated by a shortage of professional nurses; this, during a time of racial segregation and racial violence, when black Americans were very likely to be denied access to health, educational and political opportunities.

An unsuccessful campaign by black leaders failed to persuade the Army Nurse Corps, the Navy Nurse Corps or the Red Cross to admit qualified black nurses to their ranks.

Aileen Cole and Clara Rollins both wanted to become nurses; Cole, at 24, had recently passed exams for nursing and Rollins, 34, had years of nursing experience. Both lived in housing near Freedman's Hospital, in Washington, D.C., where they had graduated from the hospital's nursing training school. The Red Cross had officially enrolled both, but had done nothing else. The Red Cross was the official recruiter for the U.S. Army and Navy; neither accepted black nurses.

The worsening pandemic forced changes; the Red Cross called up Cole and Rollins and several other black nurses for duty and sent them to West Virginia to battle the pandemic. It also created opportunities for other previously excluded black nurses, including the first 18 to serve in the Army Nurse Corps between December 1918 and August 1919.

Black nurses fulfilled critical roles during and after World War I, but never have received substantial recognition for their work.

October, 1918—Philadelphia experienced a severe outbreak of the flu. The number of dead exceeded the city's morgue facilities, forcing cold storage plants to be used as makeshift morgues.

Major cities mandated closing theaters and prohibited public events. In San Francisco officials mandated that anyone outdoors must wear a mask.

Dock workers in New York were hard hit, and productivity dropped 40 percent.

November 11, 1918—Armistice Day was celebrated to mark the end of World War I. The U.S. military had mobilized over 4.3 million soldiers. Celebrations throughout the country also resulted in a resurgence of the flu.

November, 1918—Officials in Salt Lake City placed Quarantine signs on the doors of over 2,000 homes where residents had the flu.

December, 1918—Public officials began education programs to prevent the spread of the flu; the Committee of the Americas Health Association recommended stores and factories stagger openings and closings to minimize traffic.

January, 1919—San Antonio reported increasing cases of the flu. New York City reported 706 cases and 67 deaths. In February, cases in New Orleans dropped significantly.

Spring, 1919—The third wave of the flu swept through the United States; it was less deadly but still killed thousands throughout the country.

1919—Were the residents of Alaska immune because of the distance to the "lower forty-eight"? They were not. Laura Spinney tells this story in *Pale Rider: The Spanish Flu of 1918 and How it Changed the World:*

The flu entered Alaska at Unalaska Island, one of the most landward islands in the Aleutian chain that forms the tail of the Alaska Peninsula, and hence a natural stopping point for north-bound ships. The story of how it spread from there, north-east-ward toward Bristol Bay, is the stuff of legend. A Russian priest. Father Dimitri Hotovitzky—known to his flock as Father Hot Whiskey—travelled from Unalaska up to the bay to lead celebrations of Orthodox Easter, and it is said that whoever attended his services returned home sick. That he infected the bay's population is possible but unlikely. The incubation period for the flu, during which a person can be infected but symptom free, is between one and four days. Orthodox Easter fell on 20 April in 1919, coinciding as it sometime does, with "western" Easter. The first cases in Bristol Bay were reported around 12 May, three weeks later. Even allowing that some earlier cases may have gone unreported, three weeks is an unreasonably long incubation period. Someone who followed in Hot Whiskey's footsteps more like brought the virus in.

And, Spinney writes.

Close to 150 orphans were rescued from all points of the bay while the epidemic was raging. They were freezing and shivering in cold huts, without fires or food, with little clothing or bedcovering—many of them crying, huddled about their dead, reported one APA (Alaskan Packers' Association) superintendent. More were discovered after the epidemic had receded. And though the figures are unreliable, the eventual numbers of orphans that were brought to the hospital at Dillingham—a town of fewer than 200 inhabitants at the time—may have been closer to 300.

Some small native settlements, miles from anywhere, assmned they were immune from the flu because they were "in the middle of nowhere," so to speak, but when the mail came in, the flu came with it. And with the same deadly results.

The same events occurred in other outposts. In *Flu: The Story of the Great Influenza Pandemic of 1918 and the Search for the Virus That Caused It*, Gina Kolata writes:

The flu came to the islands of Western Samoa on November 7, 1918 arriving with the crew of a steamer, *Talune*, from Auckland that had docked in the tropical paradise. In the next two months, 8,000 people on the islands died of influenza, a fifth of the population.

... and the same happened in other scattered outposts of civilization, literally throughout the world.

1918-1920—Estimates of world-wide flu fatalities vary; low estimates are around 20 million with a population of 1.8 billion people at the time, approximately one percent of the population. High estimates put the death toll at 100 million, equivalent to a five precent morality rate. About 20 million died because of World War 1 in a one-year period. One in four people showed symptoms of the flu, and coimtless others had the virus but showed no symptoms. In the United States, there were approximately 675,000 deaths out of a population of 103 million at the time, a mortality rate of less than one percent.

Three questions we asked earlier: why were healthy young soldiers hard hit by the flu—why did many die within 48 hours and why were their lungs full of fluid?

Both questions are interrelated: they died of a *cytokine storm,* a physiological reaction in humans—and other animals—in which the immune system causes an uncontrollable and excessive release of pro-inflammatory signaling molecules called *cytokines.* Normally *cytokines* are part of the body's immune response to infection, but the sudden release in large quantities can cause multi-system organ failure and death. When their lungs filled, they died, drowned in their own fluids.

In short, the body's massive emergency response to the flu killed the patient.

During those years the life expectancy in America dropped by 12 years.

And further, why did doctors not really know what they were fighting, during those frightful years of 1918-1919?

The 1918-1919 flu was caused by an H1N1 virus with genes of an avian origin. This virus can only be studied effectively using an *electron microscope;* they were first developed during the 1920s and into the 1930s. The first available electron microscope in the United States was in 1937. Doctors in 1918-1919 couldn't find what they were looking for; microscopes of that period were too rudimentary, too crude; doctors literally couldn't see, or find, the H1N1 virus. And without electron microscopes to study the H1N1 virus, there was no way to develop effective vaccinations.

The Survivors

Individuals met the scourge of the 1918-1919 flu with various levels of severity and a variety of recoveries. These are, at random, some stories from those years:

Woodrow Wilson

Alfred Crosby tells this story in his *American's Forgotten Pandemic:* the key date was Thursday April 3, 1919:

> At three o'clock that afternoon President Wilson appeared to be overworked but healthy, but soon his voice began to grow so husky that Clemenceau and Lloyd George commented upon it. By six o'clock he began to cough; his cough was so convulsive as to interfere seriously with his breathing, soon he could barely walk. His temperature rose to 103 (degrees) and he was seized with the cramps of severe diarrhea. The onset of the President's illness was so abrupt and its symptoms so violent that Grayson (his personal physician) suspected poisoning. The doctor was able to control

the spasms of coughing, but the President's life remained in jeopardy throughout the night, which Grayson, who had the most important man in the world possibly dying under his hands, admitted was one of the worst he had ever spent. His diagnosis was influenza.

Crosby also writes:

A number of people who were in close contact with Wilson in spring 1919 noted an abrupt change for the worse in the man at or about the time of his attack of influenza. Lloyd George in his *Memoirs of the Peace Conference,* speaks of Wilson's "nervous and spiritual breakdowns in the middle of the Conference," and blames it on overwork and acute sensitivity to the vicious attacks of the press. Herbert Hoover, not a member of the American Peace Commission but often in Paris and in contact with the President, felt that it was Wilson's April illness which changed him.

* * *

Those who were even more intimate with the President also noted change. Gilbert Close, Wilson's secretary, wrote on April 7, "I never knew the President to be in such a difficult fame of mind as he is now. Even while lying in bed he manifested peculiarities." Ray Baker noted that he had never seen the President as worn out as in the latter part of April and May. Wilson's left eye and the whole left side of his face were twitching, drawing down the underlid of his eye. Edmund W. Sterling, the President's Secret Service guard, judged that the flu had left his chief very weak, and "he never did regain his physical strength, and his weakness of body naturally related upon his mind. He lacked his old quickness of grasp."

Crosby writes that Grayson, Wilson's physician, said that influenza was "one of the contributory causes of his final breakdown."

Woodrow Wilson suffered a major stoke in September, 1919, which forced his near-total retirement from public life. He died February 3, 1924, at 67.

JOHN STEINBECK

Influenza hit John Steinbeck while he was in high school in Salinas, California. In his massive (536 page) 1995 book, *John Steinbeck: A Biography*, Jay Parini writes:

> ... he succumbed to a flu in May of 1918 that turned quickly into pneumonia. This was the same flu that killed hundreds of of thousands of Americans that year, and it was deeply feared. Steinbeck came home from school one day "pale and dizzy" and collapsed in his bed, sending Olive (his mother) into a frenzy. His temperature shot up dangerously, and he was soon delirious. In those days before the advent of antibiotics, there was every reason for panic. "I went down and down," Steinbeck later remembered, "until the wingtips of angels brushed my eyes." Dr. Merganser, a local surgeon, was called in, and he turned the downstairs, bedroom into a makeshift operating theater. He opened the teenager's chest under ether, removing a rib to gain access to the infected lung, which was then drained of pleural pus. "We thought he would surely die on us," his sister says. "John looked horrible. We did everything we could for him. And then he had a relapse. It took a long time but he was all right in the end. I must say, we were scared to death."

And, Parini writes.

> The psychological damage inflicted by this illness of 1918, which brought him close to death and required such intrusive surgery, was considerable. It seems to have given him a sense of himself as someone on the edge of life, someone vulnerable—physically and emotionally.

Franklin Roosevelt

The S.S. Leviathan, the largest ship at that time, carried troops to Europe from the east coast. Often called the *Levi Nation* by soldiers, it was an enormous, perfect petri dish for incubating the flu virus.

Laura Spinney describes conditions on the *Levi Nation* on one Atlanta crossing. It …

> … left Hoboken, New Jersey, for France on 29 September 1918, with 9,000 military personnel and the ship's crew aboard. The disease broke out as soon as it left harbor, and by the time it docked in Brest a week later, 2,000 men were sick and there had been around 90 fatalities on board. The scenes the ship's passengers witnessed during the voyage were Dantesque. The spaces between the bunks in the troop compartments were so narrow that nurses tending the sick couldn't avoid tracking blood between them. With the higher bunks unusable by the sick, semi-conscious men were laid out on the decks instead, which soon became slippery with blood and vomit. "The conditions during the night cannot be visualized by anyone who had not actually seen them," wrote one American solider who made the crossing, and the "groans and the cries of the terrified added to the confusion of the applicants clamoring for treatment."

Franklin Roosevelt, as Undersecretary of the Navy, made a return trip from Europe on the *Levi Nation,* presumably still full of the flu virus. He had to be carried off the ship on a stretcher, although he may well have caught the flu in France, earlier. His influenza turned into double-pneumonia and he was ill for a month. Alfred Crosby writes that Eleanor Roosevelt, at one time, had Franklin, five children and three servants to care for, all down with the flu at the same time.

Roosevelt contracted what was believed to be polio in 1921; in 1932 he defeated Herbert Hoover and became the 32nd President of the United States. By common agreement (which would probably not happen today) the press did not reveal he was paralyzed from the waist down. There are almost no photographs extant showing him in heavy leg braces.

WALT DISNEY

Catherine Arnold has this paragraph in *Pandemic 1918:*

... in Chicago, a young man nicknamed "Diz" lied about his age and enlisted as an ambulance driver with the Red Cross. Ignoring his father's desperate pleas, young Diz took himself off to the Red Cross ambulance Corps training facility on Chicago's South Side and was soon learning how to drive and repair cars and trucks. When Diz himself was struck down by influenza like so many healthy young men, his father's fears had been realized. Sent home and nursed by his mother, Diz survived to become the world's most successful animator under his full name of Walt Disney.

Arthur Conan Doyle

When his oldest son died of influenza, Conan Doyle gave up writing fiction and turned to Spirituality, believing he, and others, could contact the dead.

John Dos Passos

He crossed the Atlantic on a troopship with Spanish flu on board and men dying each day. Dos Passos got influenza. He later wrote a letter home, describing something he got as a combination of: "pneumonia, T.B., diphtheria, diarrhea, dyspepsia, sore throat, whopping cough, scarlet fever and beri-beri, whatever that is," and claimed to have cured himself by self-quarantining with a bottle of Rum he bought from an English pimp.

Edvard Munch

A Norwegian artist, his most famous painting is "The Scream." Was it somehow flu-related? We can't be sure. He did have the Spanish flu and he was melancholic by nature. After he recovered he painted a series of self-portraits, one showing him yellow and gaunt in a wicker chair.

Was "The Scream" based on influenza-related delirium? He once said:

> "One evening I was walking along a path, the city on one side and the fjord below. I felt tired and ill. I stopped and looked out over the fjord—the sun was setting, and the clouds turning blood red. I sensed a scream passing through nature; it seemed to me that I heard the scream."

Others who wrestled with the flu included: General John "Black Jack" Pershing; Kaiser Wilhelm II; Georgia O'Keeffe; Franz Kafka, Mahatma Gandhi and on … and on … and on …

INFLUENZA IN AMERICA LITERATURE

There are only two American writers specifically associated with the Spanish flu. Neither Steinbeck, Hemingway, Faulkner nor F. Scott Fitzgerald nor John Dos Passos devoted much attention to the flu in their assorted books. Only two: Katherine Anne Porter and Thomas Wolfe.

KATHERINE ANNE PORTER

Porter was born in 1890; she had a variety of small jobs including being a movie extra. In 1915, she was diagnosed with tuberculosis and spent two years in sanitariums recovering; she began working as a journalist in 1917 for *The Fort Worth Critic.*

In 1918, she began working for *The Rocky Mountain News,* in Denver; she almost died the same year from influenza. It took her months to recover; when she did she was frail and bald. Her hair, formerly dark, eventually regrew but was white and remained white for the rest of her life.

Her experiences resulted in *Pale Horse, Pale Rider,* first published in 1939.

Her title is clearly from the Bible, Revelation 6:8:

> And I looked and behold a pale horse:
> and his name that sat on him
> was Death, and Hell followed with
> him. And power was given unto
> them over the fourth part of the
> earth, to kill with sword, and with

> hunger, and with death, and with
> the beasts of the earth.

There are three novellas in the book, "Old Mortality," "Noon Wine" and "Pale Horse, Pale Rider." She apparently hated the term novella and believed it had no real meaning. The last, at 67 pages, is fictional autobiography. Very thinly fictional.

Porter worked for *The Rocky Mountain News;* the protagonist, Miranda, works for a newspaper. The novella is dreamlike; Miranda meets and falls in love with a handsome man due to enter the army to be sent to Europe to fight in World War I.

Then Miranda gets influenza. When she finally recovers she is given a note. Adam, whom she loved, had died in an Army camp before being sent to Europe.

The denouement:

"Do you suppose, Mary," asked Miranda, "I could have my own room back?"

"That should be easy," said Mary. "We stored away all your things there with Miss Hobbe." Miranda wondered again at the time and trouble the living took to be helpful to the dead. But not quite dead now, she reassured herself, one foot in either world now; soon I shall cross back and be at home again. The light will seem real and I shall be glad when I hear that someone I know has escaped from death. I shall visit the escaped ones and help them dress and tell them how lucky they are, and how lucky I am still to have them. Mary will be back soon with my gloves and my walking stick, I must go now, I must begin saying good-by to Miss Tanner and Dr. Hildesheim. Adam, she said, now you need not die again, but still I wish you were here; I wish you had come back, what do you think I came back for, Adam, to be deceived like this?

At once he was beside her, invisible but urgently present, a ghost but more alive than she was, the last intolerable cheat of her heart; for knowing it was false she still clung to the lie, the unpardonable lie of her bitter desire. She said, "I love you," and stood up trembling, trying by the mere act of her will to bring him to sight before her. If I could call you up from the grave I would, she said, if I could see your ghost I would say, I believe ... "I believe," she said aloud. "Oh let me see you once more." The room was silent, empty, the shade was gone from it, struck away by the sudden violence of her rising and speaking aloud. She came to herself as if out of sleep. Oh, no, that is not the way, I must never do that, she warned herself. Miss Tanner said, "Your taxicab is waiting, my dear," and there was Mary. Ready to go.

No more war, no more plague, only the dazed silence that follows the ceasing of the heavy guns; noiseless houses with the shades drawn, empty streets, the dead cold light of tomorrow,. Now there would be time for everything.

Katherine Anne Porter eventually taught at variety of universities including: the University of Chicago; the University of Michigan; the University of Virginia; Stanford; Washington and Lee University and the University of Texas. Her only novel, *Ship of Fools*, was published in 1962; she was awarded the Pulitzer Prize in 1966. She died in 1980, at 90.

THOMAS WOLFE

Wolfe was born in Asheville, North Carolina in 1900 and eventually became known as North Carolina's greatest writer. He was one of eight children and his younger brother Ben became the focus of Wolfe's description of death by influenza.

He attended the University of North Carolina and graduated in June, 1920; in 1922 he received a master's degree from Harvard. He taught English periodically for New York University for seven years.

In 1926 he began working on a novel, first called *O Lost,* subsequently *Look Homeward, Angel,* published in 1929, and it chronicled his family and friends in Asheville, called Altamont in the novel. The Wolfe family became the Gant family and Wolfe himself became Eugene Gant. Reportedly the first version of the book was 1,100 pages; in final published form it is still a brick of a book, at 508 pages. It owes much to the style of James Joyce.

Chapter 35 details the death of his brother Ben. In *Thomas Wolfe: a Biography,* Elizabeth Nowell said that the death of Ben was "the finest writing of his career."

He wrote:

> Ben's long thin body lay three quarters covered by the bedding, its gaunt outline was bitterly twisted below the covers, in an attitude of struggle and torture. It seemed not to belong to him, it was somehow distorted and detached as if it belonged to a beheaded criminal.
>
> And the sallow yellow of his face had turned gray; out of this granite tint of death, lit by two red flags of fever, the stiff black furze of a three-day beard was growing. The beard was somehow horrible; it recalled the corrupt vitality of hair, which can grow from a rotting corpse. And Ben's thin lips were lifted, in a constant grimace of torture and strangulation, above his white somehow death-looking teeth, as inch by inch he grasped a thread of air into his lungs.
>
> And the sound of this gasping—loud, hoarse, rapid, unbelievable, filling the room, and orchestrating every moment of it—gave the scene its final note of horror.

Ben lay upon the bed below them, drenched in light, like some enormous insect on a naturalist's table, fighting, while they looked at him, to save with his poor wasted body the life that no one could save for him. It was monstrous, brutal.

* * *

The only sound in the room now was the low rattling mutter of Ben's breath. He no longer gasped; he no longer gave signs of consciousness or struggle. His eyes were almost closed; their gray flicker was dulled, coated with the sheen of insensibility and death. He lay quietly upon his back, very straight, without sign of pain, and with a curious upturned thrust of his sharp thin face. His mouth was firmly shut. Already save for the feeble mutter of his breath, he seemed to be dead—he seemed detached, no part of the ugly mechanism of that sound which came to remind them of the terrible chemistry of flesh, to mock an illusion, at all belief in the strange passage and continuance of life.

He was dead, save for the slow running down of the worn-out machine, save for that dreadful mutter within him of which he was no part. He was dead.

Thomas Wolfe wrote four memorable books: *Look Homeward, Angel; Of Time and the River; You Can't Go Home Again* and *The Web and the Rock*. He died September 15, 1938, at 37, from complications of tuberculosis.

*... a page from the Middle Ages bound
into the Twentieth Century ...*

3.

Bubonic Plague in Algeria

THE PLAGUE, ALBERT CAMUS, 1947

And, indeed, as he listened to the cries of joy rising from the town, (he) remembered that such joy is always imperiled. He knew what those jubilant crowds did not know but could have read from books; that the plague bacillus never dies or disappears for good; that it can lie dormant for years and years in furniture and linen-chests; that it bides its time in bedrooms, cellars, trunks, and bookshelves; and that perhaps the day will come when, for the bane and the enlightenment of men, it would rouse up its rats again and send them forth to die in a happy city.

Was this paragraph somewhere in Daniel Defoe's *A Journal of the Plague Year*, first published in 1722? It is, in fact, the last paragraph from *The Plague*, by Albert Camus, published 225 years later, in 1947.

There are many parallels between Defoe and Camus: Defoe reached back to recreate the Bubonic Plague of London, 1665-1666, which killed an estimated 100,000 residents of London and which occurred when he was five years old. Camus knew that the Algerian city of Oran was decimated by Bubonic Plague in 1556, and 1678, as well as minor outbreaks of the Plague there in 1921, 1931 and 1944.

Camus, by nature, was more of a philosopher than Defoe. Defoe wrote as a nascent journalist and historian; *The Plague*, by Camus is a novel; and the viewpoint is either existentialist (a term he denied) or absurdist, which he distinctly employed. In his novel, individuals are without the power to control their own destinies. In that regard, his narrative style is compatible to Kafka.

In 1948, *The New York Times* asked British poet-novelist-essayist-critic Stephen Spender to review *The Plague* for *The Times*. That review was reprinted in part, again April 15, 2020, under the title "Looking at Albert Camus's 'The Plague.'" Spender wrote:

> "The Plague is a parable and a sermon, and should be considered as such. The message is not the highest form of creative art, but it may be of such importance for our time that to dismiss it in the name of artistic criticism would be to blaspheme against the human spirit.
>
> M. Camus is a master of the Defoe-like narrative. Against the background of events, he creates various attitudes of human beings toward the plague, heightened by touches of intimate observation. As far as possible he isolates his people from their private lives, and thrusts them into their public situation.

And it can also be read as an allegory; the plight of the French under the domination of Nazi Germany, during World War Two, which Camus knew all too well.

He was born in French Algeria, of *Pieds Noir* parents, those who had lived in Europe before coming to Algeria and who presumably would return to Europe. *Pieds Noir* is a slang term for *black foot*, those French born in Algeria. He never knew his father who died in the Battle of the Marne in 1914 during World War I. He also knew disease all too well; he was diagnosed with tuberculosis when he was 14. At 20, he studied philosophy at the University of Algiers.

Camus was in Paris when the Nazis invaded France. He volunteered to join the French Army, but was refused because he had earlier suffered from tuberculosis. He was effectively quarantined when he tried to flee but finally joined the French resistance. He became editor of *Combat*, a newspaper outlawed at the time.

Like Defoe (who used many), Camus used a pseudonym to avoid being captured by the Nazis and carried fake ID cards.

Also like Defoe, he was politically active; he was by nature a leftist; he had joined the French Communist Party in early 1935 to fight unequal treatment between European and native Algerians; he left the party a year later. He subsequently joined the Algerian Communist Party, when it was founded in 1936. but was later expelled from that party because he was not sufficiently enthusiastic about party principles.

In fact, it could reasonably be said that, in terms of writing, publishing and political activism, Camus was the Defoe of the twentieth century, except for the bad debts that Defoe incurred during his lifetime.

Camus was actively leftist his entire life:

- He opposed totalitarian communism and exposed libertarian socialism;
- He supported integration in Europe and was a founder of CFFE, the French Committee for the European Federation;

- In the late 1940s, he established the *Groupes de Liaison Internationale*—the GLI—a trade union movement;
- He opposed Soviet intervention in Hungary;
- He also opposed Franco's totalitarian policies in Spain.

In Paris he had begun his "first cycle" of works—examining the meaningless and the absurd. He envisioned each cycle as including a novel, an essay and a play.

His first cycle included the novel *L'Étranger* published in Great Britain as *The Outsider* and in the United States as *The Stranger*. The essay was *Le Mythe de Sisyphe—The Myth of Sisyphus*—and the play was *Caligula*.

He moved to Algeria, then to the French Alps, to improve his tuberculosis. There he began working on his second cycle of works—the overriding subject: revolt. The novel was *La Peste—The Plague*—and the play was *Le Malentendu—The Misunderstanding*.

There are six main characters in the book and five parts. Its locale is Oran, Algeria. For readers momentarily unfamiliar with the locale and the French-Algerian names, the main characters include:

- Dr. Bernard Rieux. He treats the first victims of the Plague and like others, who observe rats in the streets, does not fully grasp the possible enormity of the plague; but quickly does realize the possibilities of the plague and warns authorities of the consequences. He believes the epidemic could kill half the population of the city, then totally 200,000. He does not believe in God, nor does he practice medicine as a noble cause; he does so simply because he is a doctor and realizes that combating the plague—death—is something he can not ultimately win.

All individuals in the novel are powerless to change their futures—their ultimate destinies. (This is the existential view which Camus denied);

- Jean Tarrou—arrived in Oran before the Plague. He is not a business man; his arrival in Oran something of a mystery. He keeps a diary which the (unknown) Narrator uses. He develops the idea of enlisting volunteers to fight the plague, rather than having city authorities conscript citizens. Tarrou becomes one of the last victims of the Plague; Camus describes his death in Part Five (to be cited later);
- Raymond Rambert, a journalist who is simply visiting Oran to research an article about the standards of living in the Arab colony in Oran. When the Plague begins, he becomes quarantined there. He struggles to escape but eventually decides to stay and help fight the Plague. He would be embarrassed if he did not;
- Joseph Grand. A 50-year old government clerk, he is privately writing a book, but can only write and re-write the first sentence and can rarely find the right words to express himself. He does however, command civic volunteers fighting the Plague and compiles statistics. He eventually catches the plague himself;
- Cottard. An eccentric, he attempts suicide but is anxious that Dr. Rieux not report it. When the Plague begins, his personality changes; he becomes agreeable and finds it easier to live with the fears he has internalized knowing that others throughout Oran face the same fears;
- Father Paneloux, a well-regarded Jesuit priest; when the Plague begins he insists in a sermon that the Plague is a punishment sent by God to those who have hardened their hearts against Him. He witnesses the death of a

child and later conducted another sermon explaining that the death of a child is a test of faith. He becomes ill but his symptoms do not indicate Plague. His death as eventually listed as a "doubtful case";
- And there is an unidentified Narrator who witnesses events during the Plague; his identity is eventually revealed in the last part of the novel.

There are least 12 or so minor characters throughout the book.

Neither Daniel Defoe nor Albert Camus ever—presumably, assuredly—witnessed a death by Bubonic Plague, but this is how Camus describes it:

> At seven Mme. Rieux returned to the bedroom. The doctor went to the surgery to ring up the hospital and arrange for a substitute. He also decided to postpone his consultations, then lay down for some moments on the surgery couch. Five minutes later he went back to the bedroom. Tarrou's face was turned toward Mme. Rieux who was sitting close beside the bed, her hands folded on her lap; in the dim light of the room she seemed no more than a darker patch of shadow. Tarrou was gazing at her so intently that, putting a finger to her lips, Mme. Rieux rose and switched off the bedside lamp. Behind the curtains the light was growing, and presently, when the sick man's face grew visible, Mme. Rieux could see his eyes still intent upon her. Bending above the bed, she smoothed out the bolster and, as she straightened up, laid her hand for a moment on his most, tangled hair. Then she heard a muffled voice, which seems to come from very far away, murmur: "Thank you," and that all was well now. By the time she was back in her chair Tarrou had shut his eyes,

and, despite the sealed mouth, a faint smile seems to hover on the wasted face.

At noon the fever reached its climax. A visceral cough racked the sick man's body and he now was spitting blood. The ganglia had ceased swelling, but they were still there, like lumps of iron embedded in the joints. Rieux decided that lancing them was impracticable. Now and then, in the intervals between bouts of fever and coughing fits, Tarrou still gazed at his friend. But soon his eyes opened less and less often and the glow that shone out from the ravaged face in the brief moments of recognition grew steadily fainter. The storm. lashing his body into convulsive movement, lit up with ever rarer flashes, and in the heart of the tempest he was slowly drifting, derelict. And now Rieux had before him only a masklike face, inert, from which the smile had gone forever. This human form, his friend's, lacerated by the spear-thrusts of the plague, consumed by searing, superhuman fires, buffeted by all the raging winds of heaven, was floundering under his eyes in the dark flood of the pestilence, and he could do nothing to avert the wreck. He could only stand, unavailing, on the shore, empty-handed and sick at heart, unarmed and helpless yet again under the onset of calamity. And thus, when the end came, the tears that blinded Rieux's eyes were tears of impotence; and he did not see Tarrou roll over, face to the wall, and die with a short hollow groan as if somewhere within him an essential chord had snapped.

The next night was not one of struggle but of silence. In the tranquil death-chamber, beside the dead body now in everyday clothing—here too, Rieux felt it brooding, that elemental peace which, when he was sitting many nights before on the terrace high above the plague, had followed the brief foray at the gates. Then, already, it had brought to his mind the silence brooding over the beds in which he had let men die. There as here it was the same solemn pause, the less that lull that follows battle; it was the silence of defeat. But there silence now enveloping his dead friend, so dense, so much akin to the nocturnal silence of the streets and of the town was set free at last, made Rieux cruelly aware that this defeat was final, the last disastrous battle that ends a war and makes peace itself an ill beyond all remedy. The doctor could not tell if Tarrou had found peace, now that all was over, but for himself he had a feeling that no peace was possible to him henceforth, any more than there can be an armistice for a mother bereaved of her son or for a man who buries his friend.

Camus eventually published five novels. Their titles, translated from the French, are:

- *A Happy Death* (written in 1936-1938 and published posthumously in 1971);
- *The Stranger*, 1842;
- *The Plague*, 1947;
- The Fall, 1956;
- *The First Man* (incomplete, published in 1994).

Camus won the Nobel Prize for Literature in 1957. At 44, he was the second youngest ever to win that award. Only Rudyard Kipling, at 42, was younger; he won the Nobel Prize for Literature in 1907.

The article, "Albert Camus novel The Plague leads surge of pestilence fiction," by Kim Willsher, published in *The Guardian* March 28, 2020, describes his death:

Albert Camus, who was born in Algeria, where *La Peste* is set, and died in 1960, aged 46, while being driven to his Provencal home by his publisher Michael Gallimard. The car, a powerful Facel-Vega, veered of the road and hit two trees. Camus was killed instantly.

The cause of the accident, on a long stretch of straight, wide road, remains a mystery and there have been numerous theories. In his book, *The Death of Camus,* the Italian author and academic Giovanni Catelli suggests the author may have bene killed by the KGB after expressing his opposition to the Soviet regime.

The Willsher *Guardian* article quotes Camus about the novel:

Albert Camus said the novel could be read on several levels and was also an allegory of the French resistance to the Pestilence of Nazism and the German occupation during the second world war.

"The inhabitants, finally freed, would never forget the difficult period that made them face the absurdness of their existence and the precariousness of the human condition. What's true of all the evils in the world is true of plague as well. It helps men to rise above themselves."

4

AIDS in America

And the Band Played On: Politics, People and the AIDS Epidemic
Randy Shilts, 1987

There are two key individuals here: actor Rock Hudson and "Patient Zero"—a name a very, very few—even in the U.S. gay community—would ever recognize—Gaetan Dugas:

> Rock Hudson, (birth name: Roy Harold Scherer, Jr.) November 17, 1925—October 2, 1985, was an American actor. One of the most popular movies stars of his time, he had a screen career spanning more than three decades. A prominent heartthrob of the Golden Age of Hollywood, he achieved stardom with his role in *Magnificent Obsession* (1954), followed by *All That Heaven Allows* (1955), and *Giant* (1956), for which he received a nomination for the Academy Award for Best Actor. Hudson also found continuing success with a string of romantic comedies co-starring Doris Day: *Pillow Talk* (1959), *Lover Come Back* (1961), and *Send Me No Flowers* (1964). During the late 1960s, his films included *Seconds* (1966), *Tobruk* (1967) and *Ice Station Zebra* (1968).

Unhappy with the films scripts he was offered, Hudson turned to television and was a hit, starting in the popular mystery series *McMillan and Wife* (1971-1977). His last role was as a guest star on the fifth season (1984-1985) of the premiere ABC soap opera *Dynasty*, until AIDS-related illness made it impossible for him to continue.

He appeared in nearly 70 films and starred in several television productions during a career that spanned more than three decades.

Although he was discreet regarding his sexual orientation, it was known by many in the film industry during his lifetime that he was gay. In 1984, Hudson was diagnosed with AIDS. The following year, he became one of the first celebrities to disclose his AIDS diagnosis. Hudson was the first major celebrity to die from an AIDS-related illness, on October 2, 1985, at 59, in Los Angeles, California.

His death was a watershed event: symbolizing the two parts of the AIDS story; pre- and post-Hudson's death (see Timeline following).

Many, many in American and elsewhere were surprised—even shocked—that a presumably heterosexual man could die of AIDS.

In the earliest days and months and years of AIDS, its was known as *gay pneumonia,* then, more ominously, as *gay cancer* and then even more ominously as *gay plague.*

Eventually, there were, and still are, four key phrases that everyone should know and understand:

1). HIV: Human Immunodeficiency Virus is a virus that attacks the body's immune system. You can get HIV from contact with infected blood, semen or vaginal fluids. Most people get HIV by having unprotected sex with someone who has HIV. Another common way is by sharing drug needles with someone who is infected with HIV. If untreated it can lead to AIDS.

Early indicators of HIV may include flu-like symptoms including: fever; headache; muscle pain or joint pain; rash; sore throat or mouth sores; swollen lymph glands; diarrhea; weight loss; cough and night sweats.

Further symptoms (Symptomatic HIV infection) may also include shingles and pneumonia.

The four stages of HIV are;

Stage 1, Infection;
Stage 2, Asymptomatic (showing no symptoms);
Stage 2, Symptomatic;
Stage 4, progression of HIV to AIDS.

2) AIDS: Acquired Immune Deficiency Syndrome is a chronic, potentially life-threatening condition caused by HIV; patients can have an HIV infection potentially for years but with no symptoms. AIDS is diagnosed when a CD4 T cell count falls below 200 or if the patient has an AIDS-defining complication, such as a serious infection or cancer.

Infections common to HIV/AIDS include:

Pneumocystis pneumonia (PCP);

Candidiasis (thrush), a common HIV-related infection, causing a white coating on a patient's month, tongue, esophagus or vagina;

Tuberculosis, a leading cause of death with people with AIDS, in emerging or resource-limited nations;

Cytomegalovirus, a common herpes virus transmitted in bodily fluids such as saliva, blood, urine and breast milk.

Complications include:

Lymphoma. This cancer starts in white blood cells; the most common early sign is painless swelling of the lymph nodes in the patient's neck, armpit or groin;

Wasting syndrome, including significant weight loss, often accompanied by diarrhea, chronic weakness and fever;

Neurological complications, including forgetfulness, depression, anxiety, difficulty walking, and may include severe dementia;

Kidney disease;

Liver disease, which is a major complication for those who also have hepatitis B or hepatitis C.

3). Kaposi's Sarcoma are tumors in the blood vessel walls. These often appears as pink or purple lesions on the skin. The darker lesions may appear to be simply blood blisters; the Sarcoma "blood blisters" are often the most obvious indictors of serious illness, whether or not the patient previously realized he or she has HIV/AIDS. It is a rare cancer.

4). AZT, azidothymidine, was originally developed as an anti-cancer drug, but it failed in tests and was put aside. When the pharmaceutical company Burroughs Wellcome began testing potential anti—AIDS drugs, they tested a newer variation of the original AZT; it seems to block the virus' activity. It was rushed to market; samples were sent to the Food and Drug Administration and the National Cancer Institute. The FDA approved AZT as the first AIDS medication March 19, 1987, in a record 20 months. Eventually studies concluded that AZT worked in a limited way, but even limited, it was seen as progress. It then cost patients $8,000. a year. Now there are a variety of HIV drugs available to keep HIV at bay and those diagnosed with HIV can live longer and relatively healthy lives with newer medications.

Timeline

The following is a general timeline of the AIDS epidemic:

Early 20th Century

Sometime early in the 20th century, Simian Immunodeficiency jumped from chimpanzees to humans, in Central Africa. This is now known as HIV-1, called zoonotic; diseases which jump from animals to mankind.

Ebola is another disease which also jumped from animals to humanity (see the Ebola chapter).

1959

A man died in the Congo—tests revealed he had HIV—the first recorded death by HIV.

1960s

HIV-2 is believed to have jumped from monkeys to humans in west Africa, Likely Guinea-Bisseau. HIV-1 arrived in America during the late 1960s. Large numbers of Haitians were working in the Congo at the time, and some likely brought the virus to the Caribbean when they returned.

Dec. 12, 1977

Dr. Grethe Rask, a Danish physician and surgeon who had spent years working in the Congo, died of pneumonia on that date. Ten years later a blood test revealed she had HIV.

1980

April 24—The CDC receives a report on Ken Horne, a gay man in San Francisco who has Kaposi's Sarcoma. He died November 30, 1981; the CDC identified him as the first American to die in the AIDS epidemic.

1981

May 18—Lawrence Mass, a gay doctor in New York city, publishes an article for *The New York Native*, a LGBT newspaper, titled "Disease Rumors Largely Unfounded." The headline was later proved false, but the article revealed that a number of gay men had been admitted to New York hospitals with severely compromised immune systems; this is the first U.S. article to mention the disease which became known as AIDS.

June 5—The CDC publishes an article describing five cases of a rare lung infection in otherwise healthy gay men in Los Angeles; two died and the other three later died.

Also on that day, Dr. Alvin Friedman-Kein reports a cluster of Kaposi's Sarcomas in gay men in New York and California. The CDC begins to receive other reports of similar cases. This CDC article is believed to be the official start of the AIDS crisis.

July, 1981—*The Bay Area Reporter*, a LGBT newspaper in San Francisco, writes about "Gay Men's Pneumonia" and urges gay men to seek treatment. The New York Times also publishes an article "Rare Cancer Seen in 41 Homosexuals." The term *gay cancer* is used beginning at this time to describe Kaposi's Sarcoma.

August 11, 1981—Playwright Larry Kramer hosts a fundraiser in his New York apartment to raise money to fund research into HIV AIDS. Kramer soon co-founds the Gay Men's Health Crisis, a community non-profit to serve the gay community during this crisis.

1982

May 11—*The New York Times* publishes an article headlined "New Homosexual Disorder Worries Health Officials," and uses the phrase Gay-Related Immune Deficiency, or GRID, which contributed to the misconception that AIDS only affects gay men.

September, 24—The CDC uses the term AIDS for the first time, defined as Acquired Immune Deficiency Syndrome, described as "A disease at least moderately predictive of a defect in cell-mediated immunity, occurring in a person with no known cause for diminished resistance to that disease."

1983

January 1—Ward 86 is established at San Francisco's General Hospital; the clinic that develops the San Francisco Model of Care, a conception that makes patients comfortable and provides resources to deal with their AIDS challenge and allows them die with dignity. This model is adopted by medical professionals and some hospitals elsewhere and sets the standard for AIDS care.

January 7—The CDC reports the first case of AIDS in women.

March 4—The CDC issues a report that states AIDS is "most prevalent among gay men with multiple sex partners, people who inject drugs, Haitians and people with hemophilia," and suggests that sexual contact and exposure to blood and blood products are the most likely vectors, or predictors, of the disease.

May 20—Dr. Francoise Barre-Sinoussia and her colleagues at France's Pasteur Institute report a retrovirus believed to be the cause of AIDS. She and one colleague eventually receive the Nobel Prize for their work.

May 25—*The New York Times* published its first front-page article about AIDS.

June 12—At a National AIDS Forum in Denver, 11 gay men take over the stage and demand that the rights of people with AIDS be protected; that they have the right to have their voices heard and make decisions about their own illnesses and also demand that the term "AIDS victim" be replaced with "people with AIDS."

September 9—The CDC rules out the possibility of transmission of AIDS by casual contact; air, food water, but misconceptions continue for years.

1984

March 1—A Study in the American Journal of Medicine examines 40 patients with Karposi's Sarcoma and traces their sexual contacts and describes an unidentified flight attendant, "Patient O," (the O standing for Outside Southern California, where the study was done), who was said to have contact with hundreds of sex partners a year. Patient O was described as a promiscuous and even a malicious gay man who single-handedly and knowingly touched off the AIDS pandemic in the United States.

Randy Shilts identifies him in his massive, 621-page book, *And the Band Played On*—and describes him as the "Typhoid Mary" of the AIDS epidemic.

April 23—The Department of Health and Human Services (HHS) announces the discovery of a retrovirus they call HTLV-III, the cause of AIDS, announce the development of a blood test and raise hopes that a vaccine could be developed in the next two years.

July 13—The CDC recommend avoiding injection drug use and reducing needle sharing as methods of preventing transmission of AIDS.

1985

March 2—The Food and Drug Administration licenses the first blood test for HIV; blood banks begin screening the nation's blood supply.

April 22—*The Normal Heart*, an autobiographical play by Larry Kramer, playwright and AIDS activist, opens off Broadway; reviews are positive.

June 20—President Ronald Reagan presents the Medal of Freedom to Mother Teresa; after the ceremony she immediately goes to George Washington University Hospital where she meets AIDS patients. She wears no gloves or face mask or any other personal protective equipment (PPE).

July 25—Rock Hudson reveals he has AIDS and media coverage increases dramatically in the next few months; it is a watershed moment.

August 27—Ryan White, a teenager who contracted AIDS through a blood transfusion, is barred from middle school in Russiaville, Indiana. The ensuing legal battle made him a national figure, highlighting the stigmas and misconceptions about how AIDS is spread.

September 17—President Ronald Reagan mentions AIDS for the first time in public. Critics inside the gay community, and some outside, believe that his slow response to the AIDS crisis was shameful. For the first two years of the Reagan administration he said nothing about AIDS and his administration did practically nothing; Reagan and his Republican cohorts, including Jerry Falwell and the Moral Majority, wanted nothing whatsoever to do with a gay disease. They all apparently viewed AIDS victims as horrific pariahs in their decent and God-fearing society.

During that time, one prominent Republican said, in public, "Gay means 'Got AIDS Yet?'"

October 2—Rock Hudson dies of an AIDS-related illness; he bequeathes $250,000. to create the American Foundation for AIDS Research.

1986

January 16—The CDC reports that 1985 saw a 89 percent increase in AIDS diagnoses and predicts that will double in 1986.

May 1—The International Committee on the Taxonomy of Viruses officially gives the name Human Immunodeficiency Virus, or HIV, to the virus that causes AIDS.

July 18—A group of minority community leaders meets with Surgeon General C. Everett Koop to voice concerns about HIV/AIDS in communities of color, unofficially founding the National Minority AIDS Council.

November—*In the Life: A Black Gay Anthology,* the first collection of writing about the AIDS epidemic, from 29 black-gay writers, is published. It does not receive much media attention or reviews, but is a watershed movement in the AIDS crisis.

1987

February—Cleve Jones creates the first panel of the AIDS Memorial Quilt, in honor of his friend, Marvin Feldman, who died of an AIDS-related illness. The quilt panel is three feet by six feet, the size of a grave plot. Subsequent panels eventually become a huge quilt, almost the size of a football field, to remember those who have died of AIDS and whose illnesses were not acknowledged by families and friends. It eventually becomes the NAMES Project.

February 4—Entertainer Liberace dies from an AIDS-related illness. Liberace long denied he was gay; his physician initially told the press and public he died of a heart attack; a week later, the actual cause was revealed.

March 12—Playwright Larry Kramer, always an acerbic AIDS activist, helps establish the AIDS Coalition to Unleash Power, or ACT UP, a lobbying non-profit to influence government officials, drug companies and others to protect those at risk of HIV/ AIDS. Their motto was "Silence = Death," and begins a campaign for more access to experimental drugs and a coordinated national AIDS campaign.

March 19—The FDA approves AZT, the first medication to treat IDS. Although it does not cure AIDS, it does slow its progress.

March 31—Ronald Reagan and French President Jacques Chirac agree that the two countries would share credit for the discovery of HIV; French researchers were first to discover the secrets of HIV, followed soon after by American scientists. There was international internecine warfare between researchers before Reagan and Chirac stepped in.

May 15—The Public Health Service adds HIV to the immigration exclusion lists. For the next 23 years, visa applicants are required to take a blood test and might be denied entry to the U.S. if they test positive for HIV.

May 31—Ronald Reagan gives his first speech about AIDS; on June 24, he creates the first Presidential Commission on AIDS. For many, it was bitterly too little, too late.

August 5—A federal judge rules that a Florida school board cannot ban three HIV-positive brothers, Ricky, Robert and Randy Ray, from attending school. The community of Arcadia, Florida responded with death threats, bomb threats and a school boycott.

August 18—The FDA approves the first human test of a vaccine against HIV.

August 28—After weeks of threats against the family of Ricky, Robert and Randy Ray, their home is burned to the ground while they were staying elsewhere. The family announced they would leave Arcadia, Florida.

October 1—The first National AIDS awareness month begins. The CDC launches a massive public relations campaign with the theme "everyone is at risk."

October 11—The NAMES project displays its AIDS Memorial Quilt on the National Mall, in Washington, D,C, for the first time. The Quilt bears the names of 1,920 people who have died of AIDS when it was first displayed. It eventually grows to over 10,000 names, making it the largest piece of community folk art in the world.

November—Randy Shilts, a reporter for *The San Francisco Chronicle,* one of the earliest newspapers to cover the HIV / AIDS epidemic, publishes *And the Band Played On,* a massive, monumental, even hypnotic, 621-page analysis from the earliest days of the HIV /AIDS through the death of Rock Hudson.

Shilts argues that the HIV/ AIDS epidemic was allowed get out of control through governmental "incompetence, apathy and discrimination against the population groups that it affected the most."

He also identifies "Patient Zero"—Gaetan Dugas, a French-Canadian flight attendant for Air Canada, from Quebec—who Shilts calls the "Typhoid Mary" of the AIDS epidemic (cited in the following pages).

May 26—The Surgeon General releases the country's first coordinated AIDS education campaign, mailing out 107 million copies of a pamphlet titled *Understanding AIDS,* in a attempt to reach every house in America; it was the largest campaign mailing in U.S. history.

November 4—President Ronald Reagan signs the first comprehensive federal AIDS bill, the Health Omnibus Programs

Extension (HOPE) Act, which established the Office of AIDS research and authorized spending federal funds for AIDS prevention, research and testing.

December 1—The World Health Organization (WHO) declares the first World AIDS Day.

1990

April 8—Ryan White, of Russiaville, Indiana, dies of an AIDS-related illness.

June 6—The International AIDS Conference convenes in San Francisco, but a number of organizations boycott it in protest of American immigration policies.

July 26—President George H.W. Bush signs the Ryan White Comprehensive AIDS Resources Emergency Act, allocating over $220 million in federal funds for care and treatment of AIDS, in its first year. The bill is the result of a bargain—in exchange for instituting "payer of last resort" programs to cover treatment for poor and uninsured people with HIV / AIDS, conservatives won the inclusion of clauses stipulating that certain funds will only be available to states that have passed harsh criminal laws against knowingly transmitting HIV.

1991

May—The Visual AIDS Artists' Caucus launches the Red Ribbon project, creating small red ribbons to raise awareness against the stigma of HIV/ AIDS. They are worn by number of attendees during the 45th Tony Awards.

November 7—Earvin "Magic" Johnson, a future Hall of Fame basketball player and one of the most famous athletes in the country, announces he tested positive for HIV and will retire immediately. He later reveals he had unprotected sex with number of women during his career. Some believe he is gay, but the diagnosis confirms that heterosexual people can also contract HIV/AIDS.

November 24—Freddie Mercury, singer for the rock group Queen dies of an AIDS-related illness. He is the first rock star to die of HIV/ AIDS. Activist groups express their gratitude that he made his diagnosis public.

1992

AIDS becomes the leading cause of death for American men ages 25-44.

April 8—Black tennis star Arthur Ashe, who broke the color bar in American tennis and who won Wimbleton, the U.S. Open and the Australian Open, announces that he is HIV-positive and states that he believes he received tainted blood during heart surgery.

April 20—The Freddie Mercury Tribute Concert for AIDS awareness is held in London. 72,000 attend.

December 13—Fifteen year-old Ricky, the oldest of the Ray brothers, dies of an AIDS-related illness.

1993

February 3—Rudolf Nureyev, a Soviet ballet dancer who defected in 1961 and who became a star with London's Royal Ballet, dies in Paris of an AIDS-related illness.

February 6—Arthur Ashe dies of an AIDS-related **illness**.

Ashe was a native of Richmond, Virginia. Richmond was, of course, the capital of the Confederacy during the Civil War. A source of civic pride in Richmond, for generations, was Monument Avenue, a boulevard of stately homes, with the largest statue anywhere, of the Confederate past. Robert E. Lee on horseback; a second, slightly smaller statue of Stonewall Jackson, one of Jefferson Davis, and a plaque honoring a Confederate Admiral (?) that no one remembers.

(The base of the Stonewall Jackson statue cited his major battles; the Robert E. Lee statue base simply said LEE.)

And Monument Avenue has a statue of Arthur Ashe holding a tennis racket, surrounded by children. In early September, 2021, after civic protests about civil rights and the racist Confederate past, a State decision was finally made to take down the Lee statue, which was done Wednesday morning. Sept. 8, 2021.

Many believe those statues should have come down decades ago; the Stonewall Jackson statue and the Davis statue should been taken down previously—the bases of both the Lee statue and the Jackson statue had been covered with graffiti, including *Black Lives Matter* in spray paint, on both bases. When the Lee statue was removed, a crowd cheered, and at least one said, "Robert E. Lee just lost his last battle." Another said, "Virginia now rejoins the Union."

The Lee statue weighted 12 tons and it was national news when it was removed. It had stood in place for more than 130 years. (Statues from the Confederacy in the Virginia Statehouse had been removed previously.)

The Arthur Ashe statue remains, untouched by graffiti, and Monument Avenue has been renamed Arthur Ashe Avenue.

President Bill Clinton gave Arthur Ashe the Presidential Medal of Freedom, posthumously.

June—President Bill Clinton establishes the Office of National AIDS Policy to coordinate federal efforts to combat HIV/AIDS.

December 14—*Philadelphia*, a major motion picture starring Tom Hanks as a gay man with AIDS and Denzel Washington as his lawyer in a anti-discrimination lawsuit, opens to rave reviews.

It was one of the first mainstream films to deal with HIV/AIDS-related illnesses as a theme. Hanks wins the Best Actor award at the 66th Academy Awards ceremonies.

1994

AIDS becomes the leading cause of death for *all* Americans ages 25-44.

June—The FDA approves the first protease inhibitor, beginning the era of Highly-Active Antiviral Therapy (HAART). Over the years, aggressive treatment for HIV/ AIDS becomes the norm.

September 22—The National Academy of Science concludes that syringe exchange programs are effective in preventing the spread of HIV / AIDS.

November—The number of total AIDS cases in the United States passes 500,000.

1995

February 23—Greg Louganis, who has been considered the greatest diver in U.S. history and who won diving events at the 1984 and 1988 Olympics, announces he is HIV-positive. He was diagnosed before the 1988 Olympic games and kept his diagnosis a secret even after coming out as gay in 1994.

1996

The number of AIDS cases in the U.S. declined for the first time since the pandemic began. AIDS ceased to be the leading cause of death for all Americans ages 25-44, although it remained as the leading cause of death among African American men in the same age bracket.

November 21—The Food and Drug Administration Modernization Act begins accelerated drug-approval programs, establishing in law a demand frequently made by gay activists.

1998

President Bill Clinton declares HIV/AIDS a "severe and ongoing health crisis," in African American and Hispanic communities after community leaders developed a "Call to Action," to address the disproportionate amount of cases in their communities. Congress allocates $156. million for the Minority AIDS initiative.

1999

Despite declining numbers in the United States, the WHO announces that AIDS has become the fourth-leading cause of death worldwide and the number one cause of death in Africa.

STATISTICS

Generally regarded as accurate, this is the eventual analysis of HIV/AIDS:

As of 2017, 28.9 to 41.5 million have died of HIV/AIDS worldwide; an estimated 36.7 million are living with HIV.

As of 2018, 700,000 have died in the United States of HIV/AIDS -related illnesses.

RANDY SHILTS

Shilts first published *And the Band Played On* in 1987. Endorsements ("blurbs") are quite, quite remarkable:

"A monumental history."
—*Washington Post Book World*

"Shilts successfully weaves comprehensive investigative reporting and commercial page-turning pacing, political intrigue, and personal tragedy into a landmark book ... Its importance cannot be overstated."
—*Publishers Weekly*

"Stunning ... an impressively researched and richly detailed narrative."
—*Time*

"It reads like a good medical sleuth story ... but it is not fiction. It is painstakingly detailed history— well documented and fairly presented."
—*Chicago Tribune*

"A lucid and stunning indictment of public policy towards the vicious disease ... A valuable work of political history."
—*Business Week*

"The most thorough, comprehensive exploration of the AIDS epidemic to date ... It is fascinating, frightening, and essential reading. It is an intimate history."
—*San Francisco Sentinel*

"An epic work of journalism on what must rank as one of the foremost catastrophes of modern history."
—*The New York Times*

"A spellbinding tale of viral terrorism, and a textbook on how institutions work—or fail to work—in the face of such a threat."
—*San Francisco Examiner*

"*And the Band Played On* will be to gay liberation what Betty Friedan was to early feminism and Rachel Carson's *Silent Spring* was to environmentalism."
—Historian Garry Wills

Any writer would be thrilled to have received such blurbs for his or her work.

The publishing record is equally remarkable: from 1987 to 2004, *And the Band Played On* sold over 700,000 copies and was translated into *seven* languages. A major paperback edition was published in 1988.

At 621 pages, it rivals Laurie Garrett's *The Coming Plague: Newly Emerging Diseases in a World Out of Balance*.

And, like Garrett's encyclopedic book, *And the Band Played On* could easily have been a two-volume set with larger type.

A film of the same name was released in 1993 with many major actors, including: Mathew Modine; Alan Alda; Phil Collins; Richard Gere; Glenne Headly; Anjelica Huston; Steve Martin; Lily Tomlin and others.

And, in archival footage: Arthur Ashe; Rock Hudson; Magic Johnson; Liberace; Freddie Mercury; newscaster Roger Mudd; Anthony Perkins; Princess Diane; Ronald Reagan; newscaster Max Robinson; Elizabeth Taylor, Ryan White and others.

A DVD version of the film is now available.

Randy Shilts was born in Davenport, Iowa and graduated from the University of Oregon with a Journalism major, where he came out as gay.

He worked briefly—and unhappily—for a gay newsmagazine, *The Advocate* in San Francisco, before being hired by *The San Francisco Chronicle* in 1981. He became the first openly gay reporter working for a major newspaper and *The Chronicle* subsequently became the first major newspaper to actively cover the HIV/AIDS story.

AIDS became a national issue that year and Shilts found his *true metier*. He then devoted his career to writing about prejudice against gays and HIV/AIDS.

His first book was *The Mayor of Castro Street: The Life and Times of Harvey Milk*. The openly gay San Francisco politician Milk, was assassinated, with San Francisco Mayor George Moscone, November 27, 1978. Although he served only briefly. Milk became a cultural icon; he was awarded the Presidential Medal of Freedom posthumously in 2009.

The Mayor of Castro Street was published at a time when even the thought of a gay politician was novel and relatively unaccepted.

And the Band Played On came next and eventually his third book. *Conduct Unbecoming: Gays and Lesbians in the US Military from Vietnam to the Persian Gulf*, was published in 1993. A fourth book, on homosexuality in the Roman Catholic Church, was uncompleted.

Although written with regular chapters, *And the Band Played On* is thoroughly episodic. Shilts moves seamlessly between events and people in San Francisco, New York, Washington, D.C., where he devotes considerable space to political wrangling about AIDS funding. And he writes about French medical researchers who were the first to discover the secrets of the HIV virus, followed closely by American scientists.

And, of course, he writes poignantly about HIV/AIDS victims dying and the anguish of their partners, families and friends

... and those trying—at times desperately—to Live with HIV/AIDS at time when medicines were not readily available.

Here is a sample of Shilts' narrative techniques, from the introductory pages of *And the Band Played On:*

> ... from 1980, when the first isolated gay men began falling ill from strange and exotic ailments, nearly five years passed before all those institutions—medicine, public health the federal and private scientific research establishments, the mass media, and the gay community's leadership—mobilized the way they should in a time of threat. The story of these first five years of AIDS in America is a drama of national failure, played out against a background of needless death.
>
> People died while Reagan administration officials ignored pleas from government scientists and did not allocate adequate funding for ADDS research until the epidemic had already spread through the country.
>
> People died while scientists did not at first devote appropriate attention to the epidemic because they perceived little prestige to be gained in studying a homosexual affliction. Even after this denial faded, people died while some scientists, most notably in the employ of the United States government, competed rather than collaborated in international research efforts, and so diverted attention and energy away from the central struggle against the disease itself.
>
> People died while public health authorities and the political leaders who guided them refused to take the tough measures necessary to curb the epidemic's spread, opting for political expediency over the public health.
>
> And people died while gay community leaders played politics with the disease, putting political dogma ahead of the preservation of human life.

> People died and nobody paid attention because the mass media did not like covering stories about homosexuals and was especially skittish about stories that involved gay sexuality. Newspapers and television stations largely avoided discussion of the disease until the death toll was too high to ignore and the casualties were no longer just the outcasts. Without the media to fulfill its role as public guardian, everyone else was left to deal—and not deal—with AIDS as they saw fit.
>
> In those early years, the federal government viewed AIDS as a budget problem, local health officials saw it as a political problem, gay leaders considered AIDS a public heath problem and the news media regarded it as a homosexual problem that wouldn't interest anybody else. Consequently, few confronted AIDS for what it was, a profoundly threatening medical crisis.

Readers can very nearly feel his anger—and outrage—in those paragraphs.

Gaetan Dugas—Patient Zero

Dugas was a French-speaking native of Quebec.

He apparently wanted to travel the world and learned English to get a job as a flight attendant for Air Canada; their rules were that all flight attendants had be bilingual. When he was not on scheduled job flights he could travel Air Canada routes free; that job perk got him free to San Francisco, Los Angeles, New York and other cites on the Air Canada route system.

He was gay—perhaps—most definitely—outrageously gay. And somewhere along the line he got HIV/AIDS.

As Shilts writes, in *And the Band Played On:*

"I am the prettiest one."

It had been the standing joke. Gaetan Dugas would walk into a gay bar, scan the crowd, and announce to his Mends, "I am the prettiest one." Usually his friends had to agree, he was right. Gaetan was the man everyone wanted, the ideal for this community, at this time and in this place. His sandy hair fell boyishly over his forehead. His mouth easily curled into an inviting smile, and his laugh could flood color into a room of black and white. He bought his clothes in the trendiest shops of Paris and London. He vacationed in Mexico and on the Caribbean beaches. Americans tumbled for his soft Quebecois accent and his sensual magnetism. There was no place that the twenty- eight-year-old airline steward would rather have rather boys fall for him than in San Francisco.

Gaetan Dugas allegedly had 250 sexual contacts a year, when he was HIV/AIDS positive, and continued that reckless behavior for years—indeed, for nearly the rest of his life.

Randy Shilts was highly awarded for his work: he was honored with the 1988 Outstanding Author Award from the American Society of Journalists and Authors; the 1990 Mather Lectureship at Harvard University and the 1993 Lifetime Achievement award from the National Lesbian and Gay Journalists' Association.

He went in for an HIV/AIDS test but delayed getting the results until he had completed reading the page proofs for *And the Band Played On.*

He was HIV /AIDS positive.

He was able to complete his third book. *Conduct Unbecoming* and was able to attend the Hollywood premier of the film version of *And the Band Played On* but a fourth book was uncompleted.

Randy Shilts died February 17, 1994 from AIDS-related complications. He was 42 years old.

Gaetan Dugas eventually returned to—or rather perhaps retreated to—his native Quebec, and died of kidney failure March 30, 1984, at 31. But not before these horrific incidents, as reported by Shilts, between Dugas and other men whom he never saw before and never saw again, in a dark cubicle in a gay bathhouse:

> It was around this time (June, 1982) that rumors began on Castro Street about a strange guy at the Eighth and Howard bathhouse, a blonde with a French accent. He would have sex with you, turn up the lights in the cubicle, and point out his Karposi's sarcoma lesions.
>
> "I've got gay cancer," he said, "I'm going to die and so are you."

In summation, Shilts wrote:

> Whether Gaetan Dugas actually was the person who brought AIDS to North American remains a question of debate and is ultimately unanswerable. The fact is that the first case in both New York City and Los Angeles could be linked to Gaetan, who himself was one of the first half-dozen or so patients on the continent, gives weight to that theory. Gaetan traveled frequently to France, the western nation where the disease was most widespread before 1980. In any event, there's no doubt that Gaetan played a key role in spreading the new virus from one end of the United States to the other.

And further, Shilts wrote, ultimately:

At one time, Gaetan had been what every man wanted from gay life; by the time he died, he had become what every man feared.

Norman Spinrad
Journals of the Plague Years

Speculative fiction can be summed in two words: *what if?*

> *What if ABC isn't ABC, but rather XYZ?*
> *What if ABC is actually CBA, or ZYX?*
> *What if? What if?*

The first sentence in the Introduction of *Journals of the Plague Years*—Spinrad added the s's to differentiate this from the Daniel Defoe's *Plague Year* of 1722—is a nice twist on a classic beginning by Dickens:

> *It was the worst of times, and it was the saddest of times,*

and, he writes …

> *so what we must remember if we are to keep our perspective as we read these journals of the Plague Years is that the people who wrote them, indeed the entire population of what was then the United States of America, and most of the world, were, by our standards, all quite mad.*

Journals of the Plague Years is the very definition of speculative fiction, and does not mention HIV/AIDS. But it should be clear to any reader the pandemic Spinrad is referring to, from looking back from his perspective, circa the year 2143.

The four main characters are:

- Walter T. Bigelow, a staunch fundamentalist Senator with an aching passion for the White House (have we heard this before?). Said to have "holy water in his veins" and who never had sex with his wife;
- Linda Levin, a free spirited 16-year old, who gets the plague and leaves her home for the wilds of San Francisco. At that time, San Francisco is blockaded: all those inside, have the virus, are considered pariahs, in quarantine; the sick, the near-dead, the dying; they cannot get out. Those outside would not dream of trying to get in.

  Eventually she discovers her system contains an antigen: her system can completely obliterate the deadly virus in any male who has unprotected sex with her. It *must* be unprotected; *meat to meat,* Spinrad says. She becomes a Earth Mother, saving, by sex, one by one, as many men as she can entice to her bed, all the men who are affected, ill, or near death from the virus;
- A mercenary condemned to death as a foot soldier in the Army of the Living Dead, and …
- A scientist, desperately working a medical solution to the plague, realizing he has only weeks to find an answer.

Writing in the website *Speculation: The Speculative in Fiction and More,* Thursday, May 21, 2015 a contributor only identified as Jesse said:

> … the novella is divided between four perspectives: a soldier, a researcher, a senator and a teenage girl.

... Spinrad leapfrogs among them in developing the overreaching story. HIV exploding in the U.S. population through rampant sexual activity, massive quarantine zones and health cards are implemented by the paranoid Christian senator Walter T. Bigelow in an attempt to keep the virus under control. His program is an initial success, the spread of the disease falters, giving Richard Bruno a chance, deep in his laboratory, to effect a cure. A cure he does find, but soon enough the virus mutates, and he's back to square one. The disease continuing to creep through the population, Linda Levin learns, at the tender young age of sixteen that she's Got It. Her parents aghast, she runs away to find a new life in the San Francisco quarantine zone and there do what she can before her time is over. But all hell breaks loose when the concerns of Bruno's laboratory and the interests of the population at large come to loggerheads. Infected soldier John Davis conscripted in his dying days to perform one last mission, the fate of HIV in the U.S. hangs in the balance.

* * *

In the end, *Journals of the Plague Years* is scathing commentary on the US political and commercial sector's interests in mass disease, particularly HIV/AIDS. The government and pharmaceuticals yet to prove their interests entirely opposite to what Spinrad outlays, the novella remains as relevant today as its did more than two and a half decades ago. ... Critically satirical in tone, sharp in style, and getting at the throat of an issue that, while it has partially diminished its threat from the 80s, remains a threat to humanity at large, the novella is as readable today as it was twenty-five years ago.

The publication of this book was, indeed, a long and winding road. Spinrad tells the story in an Afterword:

> The journey of *Journals of the Plague Years* from initial conception to this first freestanding book publication has been a long and strange one, mirroring our passage into the Plague Years themselves in some ways, foreshadowing it in others.
>
> Back around 1986, I began to realize that AIDS was going to be more than a deadly new disease. A virtually universally fatal disease that spread by sexual contact would create, in fact had already created, a baleful new existential equation between sex and death, and that could not fail to alter our psyches and our society on the most intimate and ultimate of levels.

* * *

> The children of such Plague Years would not be like thee and me. There would be social and political consequences. The economic impact of the Plague would grow and grow. Already, religious fundamentalists of a certain extremist bent were talking about AIDs as God's judgment on sinners, as a vindication of their beliefs, and this marginal viewpoint was likely to move closer to the national front and center.

Spinrad eventually took an outline to his literary agent, who promptly told him to forget it. When she decided that he wouldn't be persuaded to forget it, she advised him take the word AIDS out of the manuscript, which he did, reluctantly.

He took a long, long, detailed outline to his publishers, Lou Aronica and Shawna McCarthy, at Bantam Books, who promptly—and predictably—rejected it as a novel. But they would take the full outline as written, but with a few changes (of course), for an anthology they were publishing in 1988, *Full Spectrum*.

Finally, he wrote,

> Fortunately for me as the author, and I believe for you as the reader too, publishing perceptions have changed, and Bantam Books, the very publisher that could not see its way clear to publishing *Journals of the Plague Years* as a long book in 1987, now feels able publish it as a shorter one in 1995.

And so it was. Publication date: 1995.

See also:
The Andromeda Strain, Michael Crichton, 1969.
The Stand, Stephen King, 1978.
The End of October, Lawrence Wright, 2020.

5

Ebola in America

THE HOT ZONE, RICHARD PRESTON, 1994

In 1994, Laurie Garrett published *The Coming Plague: Newly Emerging Diseases in a World Out of Balance;* to say it is a monumental work is not an over-statement. It is impossible to conceive a more comprehensive analysis. Her Table of Contents is two full pages:

1) Machupo
   BOLIVIAN HEMORRAGIC FEVER
2) Health Transition
   THE AGE OF OPTIMISM—SETTING OUT TO ERADICATE DISEASE
3) Monkey Kidneys and the Ebbing Tides
   MARBURG VIRUS, YELLOW FEVER, AND THE BRAZILIAN MENINGITIS EPIDEMIC
4) Into the Woods
   LASSA FEVER
5) Yambuku *EBOLA*
6) The American Bicentennial
   SWINE FLU AND LEGIONNAIRES' DISEASE

7) N'zara
   LASSA, EBOLA, AND THE DEVELOPING WORLD'S ECONOMIC AND SOCIAL POLICIES
8) Revolution
   GENETIC ENGINEERING AND THE DISCOVERY OF ONCOGENES
9) Microbe Magnets
   URBAN CENTERS OF DISEASE
10) Distant Thunder
    SEXUALLY TRANSMITTED DISEASES AND INJECTING DRUG USERS
11) Hatari: Vinidogodogo
    (Danger: A Very Little Thing)
    THE ORIGINS OF AIDS
12) Feminine Hygiene
    (As Debated, Mostly, by Men)
    TOXIC SHOCK SYNDROME
13) The Revenge of the Germs,
    or Just Keep Inventing New Drugs
    DRUG-RESISTANT BACTERIA, VIRUSES, AND PARASITES
14) Thirdworldization
    THE INTERACTIONS OF POVERTY, POOR HOUSING, AND SOCIAL DESPAIR WITH DISEASE
15) All In Good Haste
    HANTAVIRUSES IN AMERICA
16) Nature and Homo sapiens
    SEAL PLAGUE, CHOLERA, GLOBAL WARMING, BIODIVERSITY, AND THE MICROBIAL SOUP

17) Searching for Solutions
PREPAREDNESS, SURVEILLANCE,
AND THE NEW UNDERSTANDING

Garrett's book is 620 pages of densely-set small type (it could have easily been a two-volume matching set, with larger type) and 104 pages of Notes, in equally small type.

Readers would be well advised to use a magnifying glass to wade through this fascinating, but long march.

In a Preface, Jonathan M. Mann, Director of the International AIDS Center at Harvard, wrote:

> We always want to believe that history happened only to "them," "in the past." And that somehow we are outside history, rather than enmeshed within it. Many aspects of history are unanticipated and unforeseen, predictable only in retrospect: the fall of the Berlin Wall is a single recent example. Yet in one vital area, the emergence and spread of new infectious diseases, we can already predict the future—and it is threatening and dangerous to us all.
>
> The history of our time will be marked by recurrent eruptions of newly discovered diseases (most recently, hantavirus in the American West); epidemics of diseases migrating to new areas (for example, cholera in Latin America); diseases which became important through human technologies (as certain menstrual tampons favored toxic shock syndrome and water cooling towers provided an opportunity for Legionnaires' Disease); and diseases which spring from insects and animals to humans, through man-made disruptions in local habitats.
>
> To some extent, each of these processes has been occurring throughout history. What is new, however, is the increased

potential that at least some of these diseases will generate large-scale, even world-wide epidemics, (pp. xi)

The same year Laurie Garrett's massive encyclopedia was published, Richard Preston published *The Hot Zone;* it presents events between 1967 and 1993.

Preston's narrative begins with a portrait of Charles Monet (not his real name, Preston admits) who is something of a loner and who lives on a plantation on the private lands of the Nzoia Sugar Factory, within eyesight of Mount Elgon, a huge extinct volcano, in western Kenya. Monet, was, Preston says, 56, and had few close friends. Monet was in charge of the pumping system that fed water to many miles of sugarcane fields.

On New Years Day, 1980, Monet and a female friend took a Land Rover and explored Kitum Cave, a huge cave, a haven for elephants at night, full of bats and bat guano. It is the dawn of time brought to life; including a petrified rain forest inside the cave, the bones of crocodiles, ancient hippos and the mummified corpses of baby elephants who got into the cave and fell into a crevice.

Kitum Cave was, in parts, fascinating and haunting. And dangerous—there were tree trunks deep inside turned to stone, with spikes as sharp as h*yp*odermic needles.

Seven days later, January 8, 1980, Monet became ill. He had pain behind his eyeballs, which got worse; he developed a fever and began to vomit. He became passive, his face an expressionless mask. His eyeballs turned bright red, his skin turned yellowish, with "brilliant starlike red speckles." He became "sullen, resentful, angry" and appeared to have no memory. He could answer questions but seemed not to know exactly where he was, Preston writes.

When he didn't show up for work, co-workers found him in his bungalow and decided he should go to a hospital. They took him to a private hospital in Kisumu, on the shores of Lake Victoria. Doctors there thought he might have malaria and gave him antibiotics which had no effect on his illness. The doctors then thought he should go to Nairobi Hospital, the best private hospital in Africa.

He was able to travel by himself and got to the nearest airport and booked a Kenya Airways flight. The aircraft was a Fokker Friendship plane, a prop-plane which held little more than 30 passengers.

As Preston tells the story of Monet enroute to Nairobi:

> He was holding an airsickness bag over his mouth. He coughs a deep cough and regurgitates something in the bag. The bag swells up. Perhaps he glances around, and then you see that his lips are smeared with something slippery and red, mixed with black specks, as if he had been chewing coffee grinds. His eyes are the color of rubies, and his face is an expressionless mass of bruises. The red spots, which few days before had started out as starlike speckles, have expanded and merged into huge, spontaneous purple shadows: his whole head is turning black and blue. The muscles of his face droop. The connective tissue in his face is dissolving, and his face appears to hang from the underlying bone, as if the face is detaching itself from the skull. He opens his mouth and gasps into the bag, and the vomiting goes on endlessly. It will not stop, and he keeps in bringing up liquid, long after his stomach should have been empty. The airsickness bag fills up to the brim with a substance known as the *vomito negro,* or the black vomit. The black vomit is not really black; it is a speckled liquid of two colors, black and red, a stew of tarry granules mixed with fresh red arterial blood. It is loaded with virus. It is highly infective, lethally hot, a liquid that would scare the daylights

out of a military biohazard specialist. The smell of the *vomito negro* fills the passenger cabin. The airsickness bag is brimming with black vomit, so Monet closes the bag and rolls up the top. The bag is bulging and softening, threatening to leak, and he hands it to a flight attendant.

When a hot virus multiplies in a host, it can saturate the body with virus particles from the brain to the skin. The military experts then say the virus has undergone "extreme amplification." This is not something like the common cold. By the time an extreme amplification peaks out, an eyedropper of the victim's blood may contain a hundred million particles of virus. During the process, the body is partially transformed into virus particles. In other words, the host is possessed by a life form that is attempting to convert the host into *itself*. The transformation is not entirely successful, however, and the end result is a great deal of liquefying flesh mixed with virus, a kind of biological accident. Extreme amplification has occurred in Monet, and the sign of it is the black vomit.

He appears to be holding himself rigid, as if any movement would rupture something inside him. His blood is clotting up—his bloodstream is throwing clots, and the clots are lodging everywhere,. His liver, kidneys, lungs, hands, feet and head are becoming jammed with blood clots. In effect, he is having a stroke through the whole body. Clots are accumulating in his intestinal muscles, cutting off the blood supply to his intestines. The intestinal muscles are beginning to die and the intestines are starting to go slack. He doesn't seem fully aware of pain any longer because the blood clots lodged in his brain are cutting off blood flow. His personality is being wiped away by brain damage. This is called depersonalization, in which the liveliness and details of character seem to vanish. He is becoming an automaton. Tiny spots in his brain are liquefying. The higher functions of consciousness are winking out first, leaving the deeper parts of the brain stem (the primitive rat brain, the lizard brain) still alive and functioning. It could be said that the *who* of Charles Monet has already died while the *what* of Charles Monet continues to live.

Are you scared yet?

One of the people asked to contribute a blurb for *The Hot Zone* was horror novelist Stephen King. He wrote:

> The first chapter of *The Hot Zone* is one of the most horrifying things I have read in my whole life ... and then it gets worse. That's what I keep marveling over: *it keeps getting worse*. What a remarkable piece of work. I devoured it in two or three sittings, and have a feeling the memories will linger a long time.

It would take a very substantial subject and text to scare the bejesus out of Stephen King, himself the absolute master of horror novels.

Charles Monet died of his horrific—and then mysterious—illness. Did he touch something inside Kitum Cave? Worse yet, did he simply inhale something ... ?

Marburg, Germany, 1967

There has been thriving business in sales of monkeys for research purposes for years, from Africa, often through the Philippines, to Europe and also into the United States.

And viruses are sometimes named for the place of origin, a city, region or country.

The Behring Works, in Marburg, Germany, produced vaccines using kidney cells from African green monkeys, imported from Uganda. The company had received shipments of 500- 600 monkeys; only perhaps two or three of them were incubating the then-unknown virus. But soon, it spread among the monkeys; a few got sick, bled internally and died.

Then, unaccountably, the virus jumped to the human population of Marburg. The first person infected was a man only identified as Klaus F., at employee at Behring who fed the monkeys and cleaned their cages. He had notable signs of the virus August 8,1967 and died two weeks later.

Other victims, also at Behring, developed headaches, raging fevers, blood clotting, spurts of blood and terminal shock.

31 people eventually caught the virus; seven died in pools of blood. The kill rate, Preston writes was one in four, making it an extremely lethal virus. By contrast he says, yellow fever, considered a highly lethal virus, kills only about one in 20 patients.

Marburg, as it is called now, is one of a family of viruses known as *filoviruses;* it was the first to be discovered. That word in Latin means *thread virus,* so called because they are matching and do not resemble any other virus known to mankind. When they appear in a "great flooding mass," Preston says, as they do when they have killed a victim, they look "like a tub of spaghetti that has been dumped on the floor."

Effects on the human brain are similar to rabies; but since the virus is a variant, they first decided to call it a sort of "stretched rabies."

But it soon became clear that the filoviruses called Marburg was one of three variants of Ebola; the others were Ebola Zaire and Ebola Sudan.

Ebola Zaire killed nine out of 10 victims; 90 percent of people with Ebola Zaire died of it, Preston writes.

And Ebola Marburg—the gentle sister of Ebola Zaire, if such a comparison can be made—acts like nuclear radiation, damaging virtually all internal organs, intestines and skin.

Soon after the Behring outbreak the World Health Organization sent investigators to Uganda to discover where the outbreak originated. The monkeys had been trapped at locations all over Uganda, so a single source couldn't be found.

The mystery lay unsolved until 1982, when a possible source was discovered; a British veterinarian who had been working temporarily inspecting monkeys for shipment in Entebbe, where monkeys had been sent to Behring. The company he had been working for in Entebbe was exporting 13,000 monkeys a year to Europe, a substantial business.

This British veterinarian, who preferred to be anonymous in Preston's book, had seen some sick monkeys and pulled them off shipments, to be relocated to a small island in Lake Victoria, which essentially became a breeding ground for the virus, "a hot island, a land of plagues." But some of those monkeys were caught again and then shipped to Europe.

And, this man had heard, that between 1962 and 1965, when he was stationed in eastern Uganda on the slopes of Mount Elgon inspecting cattle, local tribal chiefs told him of an outbreak of a disease that caused bleeding, a skin rash—and death.

This man was never able to confirm the tribal rumors, but it seems possible that an earlier outbreak of the virus occurred on Mount Elgon.

Fort Detrick, Frederick, Maryland, September, 1983

The long name is: the United States Army Medical Research Institute of Infections Diseases, or USAMRIID, housed at Fort Detrick. It is the largest employer in Frederick County, Maryland.

Those in the military pronounce it *You-Sam-Rid* and it is the nation's defense against biological weapons. *You Sam rid* specializes in biocontainment and vaccines. Research has included anthrax; botulism and any sort of biological weapon American military forces might meet on the battlefield.

During the Second World War, labs there focused on offensive biological weapons that could be dropped from aircraft onto enemy troops. In 1969, President Richard Nixon signed an executive order which outlawed offensive biological weapons research; since that time, *You Sam rid* has been devoted to peaceful solutions to vexing and seemingly unsolvable biological problems.

The Institute "knows ways to stop a monster virus before it ignites an explosive chain of lethal transmissions in the human race," Richard Preston writes.

One of the civilian staff members working there was Eugene Johnson, an expert in Ebola and the sister virus, the Marburg strain. He had infected several monkeys with Ebola and gave them various drugs to see if the Ebola virus could be stopped.

They soon began to die.

Two of the key people in this medical/military research were Jerry and Nancy Jaax; both career Army officers, both veterinarians, *doggie docs*. Nancy Jaax, then a Major, joined the project as a pathologist; it was her job to determine the cause of death in the research monkeys.

She and Johnson inspected the research monkeys. She found two; they had bloody noses, their eyes half open, glassy and brilliant red with dilated pupils. They had no facial expressions; not even of pain. Connective tissue under the skin has been destroyed by the virus, and parts of the brain which controlled facial expression had been destroyed.

The masklike face, the red eyes and the bloody nose were classic signs of Ebola that appear in both primates and humans infected with that disease. The classic Ebola face made the monkeys look as if they seen something beyond comprehension.

It was not a vision of heaven, Preston writes.

He explains viruses and specifically Ebola succinctly in less than two pages:

> Virus are ambiguously alone, neither alive nor dead. They carry on their existence in the borderlands between life and non-life. Viruses that are outside cells merely sit there; nothing happens. They are dead. ... (but) once the virus enters a cell, it switches on and begins to replicate.
>
> A virus is a parasite. It can't live on its own. It can only make copies of itself inside a cell using the cell's materials ...

And, he writes, the genetic code inside Ebola is a single strand of RNA. This type of molecule is thought to be the oldest and most "primitive" coding mechanism for life. The earth's primordial ocean, which came into extinction not long after the earth was formed, about four and one half billion years ago, may well have contained microscopic life forms based on RNA. This suggests that Ebola is an ancient kind of life, perhaps nearly as old as the earth itself. Another hint that Ebola is extremely ancient is the way in which it can seem neither quite alive nor quite unalive.

And, ominously: the virus is dedicated to making copies of itself—which it can do on occasion with radiant speed, he said.

The research lab had control monkeys; some who had not been infected with Ebola. They were expected to remain healthy.

But ... as novelist Stephen King wrote in his blurb for Richard Preston's book:

It keeps getting worse.

The control monkeys, which had no contact with the infected monkeys, also showed signs of Ebola. *It had apparently spread through the air.*

Now what?

All research scientists, and officer/veterinarians like Major Nancy Jaax, had to wear sealed suits, with self- contained breathing units. Any haphazard cut, no matter how small, might lead to Ebola—and a gruesome death.

And the fear was—if Ebola was airborne, and if it escaped the research units at *You-Sam-Rid*, it could infect all of Fort Detrick, then all of Frederick County and then all of . . . there would be no way of stopping an airborne Ebola plague.

Army top brass in *You-Sam-Rid* were called in, as well as top people in the Center for Disease Control, the CDC, in Atlanta. After brief spats of internecine warfare between the two sides, the Army retained control of the research labs at *You-Sam-Rid*.

* * *

Roughly the last half of *The Hot Zone* is a fascinating, even hypnotic read; how the scientists and researchers and even the top brass at *You-Sam-Rid* confronted the idea that Ebola was running rampant—and airborne—among the research monkeys—killing some each day. Two men became ill and it was thought that both had Ebola; neither did and both recovered.

Does Ebola still exist in the world? Surely so. Perhaps now in the depths of Kitum Cave, among the crocodile remains, the hippo remains, the petrified rain forest and the mummified baby elephants. In those cave depths, which surely date to pre-history, it is "neither quite alive nor quite unalive."

… and may emerge again.

6

Covid-19

THE PLAGUE YEAR:
AMERICA IN THE TIME OF COVID
LAWRENCE WRIGHT, 2021

The Covid-19 pandemic may well have been a human tragedy of biblical proportions in America in 2020-2021.

This Timeline alone shows how horrific this pandemic was—and is. (These are only a fraction of the events, announcements, government actions and deaths during these months.)

December 31, 2019—The U.S. Centers for Disease Control and Prevention (CDC), becomes aware of cases of SARS-CoV-2 in China which may have been introduced into the United States prior to January 19, 2020.

JANUARY, 2020

January 3, 2020—CDC Director Robert Redfield was informed by a counterpart in China that a "mysterious respiratory illness was spreading in Wuhan (China)." He notified Health and Human Services (HHS) Secretary Alex Azan, who shared the report with the National Security Council (NSC). According to *The Washington Post,* warnings about the virus were placed in the President's Daily Brief (PDB), in early January. (President Trump was notorious for ignoring his PDBs.)

On the same day, Redfield emailed and spoke to George Gao, director of the Chinese CDC, and the next day offered technical assistance.

January 5—The World Health Organization (WHO) reported a "pneumonia of unknown cause" in Wuhan, The WHO advised against travel or trade restrictions at this time.

January 6—The CDC issued a travel notice for Wuhan and CDC Director Redfield offered to send a team of CDC scientists to Wuhan. China did not accept the offer for several weeks, which delayed the U.S. access fo the virus. China did however release genetic test data on the new virus on January 9.

During the week of January 9, officials in the HHS developed an inter-agency task force including Redfield (CDC) and Anthony Fauci, Director of the National Institute of Allergy and Infectious Diseases.

January 7—The CDC established a coronavirus incident task force management system to better share and respond to information about the virus.

January 8—The CDC issues its first public alert about the coronavirus.

January 9—The WHO issues a statement naming the new disease as a coronavirus in Wuhan.

January 10—The WHO issues a comprehensive package of guidance to countries on how to test for potential cases and warned of person-to-person risk of transmission.

January 11—The CDC updated the January 6 Level 1 travel health notice for Wuhan.

January 14—The WHO held a press briefing that stated that their information suggested a possibility of limited, but not sustained human-to-human transmission. The WHO recommended countries take precautions due to human-to-human transmission during earlier SARS and MERS outbreaks. The WHO stated that Chinese officials found no clear evidence of human-to-human transmission of the new coronavirus. The head of China's National Health

Commission, Ma Xiaowei, confidentially provided a memo which presented a "grim" situation to Chinese officials. The Chinese public was not informed about this until January 20.

January 17—The CDC dispatched public health officials to screen incoming airport passengers at John F. Kennedy Airport in New York City, as well as San Francisco and Los Angeles and monitored traffic in Chicago and Atlanta in late January.

January 18—Alex Azar discussed the coronavirus outbreak with President Trump, who accused him of being "alarmist."

January 20—Chinese authorities announced the confirmation of human-to-human transmission of the coronavirus; on the same day the first U.S. case was reported, a 35-year old American who had returned from Wuhan to his home state of Washington. The CDC developed its own coronavirus test but subsequently it was found to be defective. The CDC directed that all state health departments send all samples to the CDC in Atlanta, to speed turn-around times for testing.

Also on January 20, Chinese Communist Party General Secretary Xi Jinping and State Council premiere Li Keqiang issued the first public warning about the coronavirus to the Chinese people.

On the same day, Anthony Fauci announces that the National Institute of Health was working on the development of a vaccine for the coronavirus. State department officials and local health officials in collaboration with the CDC, began identifying all persons who had close contacts with Covid-19 patients.

January 21—The CDC activates its emergency operations center to help support the response to the coronavirus. The man who had returned from Wuhan who was hospitalized was released two weeks later after treatment; a second case was reported in Chicago and a third case reported in Orange County, California.

In a briefing that same day, CDC and Washington State officials summarized the case of the man who had returned from Wuhan, and that, so far, 1,200 passengers had been screened with negative findings. Passengers from Wuhan would be funneled into designated airports for more effective screenings.

January 22—Trump has his first question from the press if he is concerned about the coronavirus. "No, not at all. And we have it totally under control. It's one person coming in from China . . . It's going to be just fine." He subsequently said "We do have a plan and we think it is going to be handled very well. We're already handled it very well. The CDC is terrific. Very professional."

January 23—Chinese authorities lock down the city of Wuhan, which has a population of 11 million. By January 30, the entire province of Hubei was locked down.

Also on that date the WHO issued a statement, which read, in part: "All countries should be prepared for containment, including active surveillance, early detection, isolation and case management, contact tracing and prevention of onward spread of 2019-nCoV infection and to share full data with the WHO."

Also on that date, the CDC sought emergency use authorization from the FDA enabling states to use its newly developed coronavirus test.

January 24—The U.S. Senate was briefed on the coronavirus by key health officials. Four U.S. senators, Richard Burr, Kelly Loeffler, Dianne Feinstein and James Inhofe allegedly sold stock thereafter, prior to significant declines in the stock market. In Kelly Loeffler's case, stock sales took place on the same day. All denied wrongdoing. (Kelly Loeffler subsequently lost her bid for re-election in Georgia.) President Trump praised China's efforts in a tweet.

January 26—Two more cases were confirmed; both had recently returned from Wuhan.

January 27—The WHO assessed the risk of the coronavirus to be "high at the global level." Also on that date, the CDC reported that 2,400 travelers in from Wuhan has ben screened; 110 were under investigation, five had tested positive and 32 negative.

January 29—The U.S. announces a White House Coronavirus Task Force, including various personnel at the HHS, CDC and the State Department. Initially the focus was solely to keep the logistics of travelers returning from China, not on testing or supplies.

Also on that date, the government evacuated 195 State Department employees from Wuhan. *The New York Times* reported that, in a memo by Peter Navarro, the coronavirus could cause as many as one half million deaths.

January 30—HHS Secretary Azar warned President Trump of a "possibility of a pandemic."

The WHO issued a statement warning, "All countries should be prepared for containment, including active surveillance, early detection, isolation and case management, contact tracing and prevention of onwards spread." The U.S. federal government did not issue guidelines for social distancing until March 19. As late as April 8, five states had no social distancing rules and three others had rules for only parts of the states.

January 31—One more person returning from Wuhan was confirmed positive in California. All flights from China were redirected to designated airports, where public health resources had been allocated.

Donald Trump claimed credit for acting early with the travel ban, but it really wasn't a ban. *The Washington Post* reported that 300,000 people had traveled from China to the U.S. during a 14-day waiting period. It was a sieve. *The New York Times* subsequently reported that about 430,000 people traveled from China to the U.S. between December 31, 2019 and April 4, 2020. Flights from Europe weren't banned until March 11; hundreds of thousands crossed the Atlantic into the U.S. during that period, many presumably bringing the coronavirus with them.

FEBRUARY, 2020

February 3—49 member of Congress sign a letter to CDC Director Redfield highlighting the urgency for distributing a rapid diagnostic kit that could be processed locally, rather than sending to the CDC in Atlanta, which they referred to as an "unsustainable bottleneck."

February 4—The FDA issues an emergency use authorization for the CDC diagnostic kit to test for coronavirus.

February 5—A fifth case is discovered, a college student from Wisconsin. A government flight takes 345 passengers from Hubei Province, and another flight on February 6, takes 300 additional passengers from China.

HHS Secretary Azar, and others including Anthony Fauci, brief Congress on the governmental response efforts.

February 6—The CDC sends out 90 viral detectors to state-rim labs, which discovered the tests were inadequate and had to be shipped to the CDC in Atlanta instead. The Director of the WHO said, "We have shipped 250,000 tests to more than 70 laboratories around the world and we are training lab works in how to use them."

February 6—Patricia Dowd, 57, of San Jose dies. An autopsy determines that her death is the first Covid-related death in the United States.

February 7—In a recorded interview later made public by journalist Bob Woodward, Trump admits "It's also more deadly than even your strenuous flu." But for the next month he says publicly that the threat is minimal. "I always wanted to play it down. I still like playing it down, because I don't want to create a panic."

February 9—The White House Coronavirus Task Force holds a briefing with governors from across the nation attending the National Governors' Association meeting in Washington, D.C.

February 11—WHO Director-General Tedros Adhanom Ghebreyesus said, "(A) virus is more powerful in creating political, economic and social upheaval than any terrorist attack … If the world doesn't want to wake up and consider this enemy virus Public Enemy Number 1, I don't think we will learn our lessons."

February 14—The CDC announced it is working with local health departments to aid in early detection, of Covid cases.

February 15—338 U.S. nationals are evacuated from the cruise ship *Diamond Princess*, which had been held in quarantine in Yokohama, Japan. 14 tested positive for the Coronavirus; six more positive cases were later reported and then an additional six later tested positive.

February 18—The HHS announced it would work with the French firm Sanofi Pasteur to develop a vaccine against the virus.

The New York Times reported that the Chief Medical Officer of HHS, Dr. Duane C. Caneva, had conducted a series of e-mail chains begun in January under the code name *Red Dawn*, from a film of the same name; the conclusions were that gatherings on the *Diamond Princess* cruise ship were not markedly different from malls, schools or work environments; that it would be difficult for local officials to direct closings without federal help. The group concluded that "the United States had already lost the fight to contain the virus," based on the realization that many in the U.S. were already infected and capable of spreading the disease without showing any symptoms. The term NPI came into use: "Non-pharmaceutical interventions"; school, business and community closings to block the spreads of the virus.

February 20 and 21—Two more cases were reported in California; the first cases of "community transmission," because the origins were unknown.

February 22—A U.N./WHO team of international specialists arrives in Wuhan, with individuals from the U.S., Germany, Japan, Nigeria, Russia, Singapore and South Korea involved.

February 23—The CDC announced that airport screening totaled 46,016 travelers. However, only 22 were referred to hospitals; only one tested positive for Covid. That individual was hospitalized and treated. The CDC later admitted that of 700,000 screenings, temperature and symptom-based analysis was ineffective based on a variety of factors including; the relatively long incubation period; nonspecific symptoms common to other infections and travelers attempting to hide their symptoms with over-the-counter medications such as cough syrup.

February 24—Donald John Trump sent out a tweet: "The Coronavirus is very much under control in the USA ... CDC and World Health (Organization) have been working hard and very smart."

February 25—HHS Secretary Alex Azar testified before the U.S. Senate. The *National Geographic* summarized his testimony: "The Strategic National Stockpile has just 30 million surgical masks and 12 million (N95) respirators in reserve." An additional 300 million of each could be required to protect health workers. HHS said it intended to buy as many as a half billion respirators and surgical face masks over the next year and one half. The *National Geographic* stated that the "U.S. has only a fraction of the medical supplies it needs to combat coronavirus."

February 25—Dr. Nancy Messonnier, Director of the CDC's National Center for Immunization and Respiratory Diseases delivered a briefing that said, "Disruption to everyday life might be severe." *The New York Times* reported that Trump was "furious." Azar attempted to downplay her comments in a news conference later that same day.

February 26—In a news conference, President Trump said, "When you have 15 people, and the 15 within a couple of days is going be down to zero, that's a pretty good job we've done."

February 26—Vice President Mike Pence was appointed to lead the Coronavirus Task Force, replacing HHS Secretary Azar as leader. Pence was the first leader in the Trump presidency to coordinate the planning and response, two months after the government became aware of the coronavirus problem.

February 29—The first death was reported in the EvergreenHealth Medical Center in Kirkland, Washington. Governor Jay Inslee declared a state of emergency for the State of Washington.

March, 2020

March 2—Cases in the U.S. reach 100.

March 4—The CDC lifts federal restrictions on coronavirus testing, to allow any American to be tested for coronavirus "subject to doctor's orders."

California Governor Gavin Newsom declares a state of emergency.

HHS announced the intention to purchase 50 million N95 respirators in the next 18 months.

Secretary Azar announces that HHS was transferring $35 million to the CDC to help state and local jurisdictions impacted by the virus.

March 5—Hawaii, Utah, Nebraska, Kentucky, Indiana, Connecticut, South Carolina, Pennsylvania and Oklahoma all announce new cases in their states, associated with passengers from the *Grand Princess*.

Four deaths are reported in Washington, related to patients in the Kirkland, Washington, health care facility.

President Trump signs a $8.3 billion bill that day, providing 7.76 billion to federal, state and local agencies for combating the virus and authorizing an additional $500 million in waivers for Medicare Telehealth restrictions.

March 7—Virginia, Kansas, Missouri and Washington, D.C. all announced their first cases.

March 8—Iowa and Vermont report their first cases.

March 9—Speaking to the press corps in a White House Coronavirus Task Force briefing. President Trump says the virus is "very much under control" and was less deadly then the 1918-1919 flu. He predicts that the case count will soon drop to zero.

CNN declares Covid-19 a pandemic.

Ohio Governor Mike DeWine declares a state of emergency in Ohio. Ohio State University, one of the top three largest universities in the nation, suspends face-to-face classes until March 30.

March 10—South Dakota and Michigan report their first cases. Governor Charlie Baker of Massachusetts declares a state emergency. Harvard University orders its students to vacate the campus by Sunday, March 15.

President Trump and Vice President Pence meet with top health insurance companies and secure an agreement to waive co-pays for Covid-19 tests and treatment.

March 11—Arkansas, Delaware, Mississippi, New Mexico, North Dakota and Wyoming report their first cases. Ohio Governor Mike DeWine orders all gatherings of more than 1,000 people be banned in Ohio.

The University of Notre Dame suspends all in-person classes until at least April 13.

The University of Maine announces that all in-person classes will be canceled for the rest of the semester; the University of Minnesota suspends all in-person classes at least until April 1. The Mayo Clinic, in Rochester, Minnesota, begins "drive-through" testing, though it has to be approved by telephone screening.

Washington University in St. Louis moves to online classes and the University of Missouri changes from fulltime in-person classes to partial on-line classes.

President Trump says, in an Oval Office address, for "the vast majority of Americans, the risk is very, very low."

March 12—Total U.S. cases pass 1,500. Sports leagues such as Major League Soccer, the National Hockey League and the National Lacrosse League suspend their seasons in progress at that time. Major League Baseball cancelled the rest of their spring training games.

The University of Alabama system and Auburn University move to online classes after their spring break. Governor John Carney of Delaware declares a state emergency. The University of Hawaii moves classes online beginning March 23.

HHS places its first order—for 4.8 million—for N95 respirators. But the supplier could only begin delivery about the end of April. Former HHS Secretary Kathleen Sebelius said "We basically wasted two months."

March 13—Total U.S. cases passed 2,100.

HHS issues a report "not for public distribution" that stated "the pandemic will last 18 months or longer and could include multiple waves of illness" and that the then-current supply-and-transportation chain "would likely result in significant shortages."

President Trump issues a declaration stating a "national state of emergency."

His declaration enabled FEMA to mobilize existing personnel to respond more quickly to help local and state agencies to deal with the pandemic.

Governors of six states declare a states of emergency in their respective states: Brian Kemp of Georgia; Tim Walz of Minnesota, Christopher Sununu of New Hampshire; Kate Brown of Oregon; Henry McMaster of South Carolina and Kristi Noem of South Dakota.

March 13—Total U.S. deaths now exceed 2,700.

The CDC issues guidance against any gathering of more than 50 people. Governors Janet Mills of Maine and Kevin Stitt of Oklahoma declare states of emergency. Schools continue to close through the country.

March 16—President Trump issues new guidelines urging citizens to avoid gathering of more than 10 people, but stops short of ordering quarantine or curfew, but does say the country is headed toward a recession. All bars, movie theaters, gyms and fitness centers are ordered closed in Los Angeles and restaurants are permitted take-out orders only; these rules extend largely throughout the country.

March 17—Governor Laura Kelly of Kansas orders all K-12 schools closed for the rest of the school year; deaths continue to climb throughout the nation. Governor Mike DeWine of Ohio orders all elective surgeries in Ohio hospitals be postponed; other states also cancel elective operations, on different timelines.

President Trump *was* quoted as saying "This is a pandemic ... I felt it was a pandemic long before it was called a pandemic."

Secretary of Agriculture Sonny Perdue announces a partnership between USDA, Baylor University, McLane Global and Pepsi Co. to provide one million meals per week to rural children in response to widespread school closings.

The Treasure Department deferred $300 billion in tax payments for 90 days without penalty, up to one million for individual and 10 million for business.

March 18—U.S. Congressman Mario Diaz-Balart is one of the first members of Congress to test positive for the coronavirus, as does Congressman Ben McAdams of Utah.

Reverend Tony Spell of East Baton Rouge, Louisiana, hosts hundreds at a service in his Life Tabernacle Church in open defiance of Governor Jon Bell Edwards' ban against gatherings of more than 50 people.

March 19—3,000 individuals sign a letter asking ICE to release individuals and families detained for immigration violations, as overcrowding condition are ripe for spreading the coronavirus.

Two cruise ships are prevented from disembarking in Hawaii, despite having no cases of Covid-19 onboard.

ProPublica reveals that Senator Richard Burr of North Carolina allegedly used his position as chairman of the Senate Intelligence Committee to mislead the public about Covid-19. He personally made between $582,029 and $1.56 million by selling off stock days before the market crashed.

Stay-at-home orders are issued throughout the country.

March 20—North Carolina activates the state National Guard to assist in logistics and transportation of medical supplies; the state has 179 cases as of that date.

March 21—Florida Governor Ron DiSantis is considering a new strategy to put positive Covid-19 patients in isolation shelters such as abandoned convention centers and hotels instead of returning the patients to their homes where they can infect their own families and others.

Throughout the rest of the Trump administration, and into the Biden administration, Florida Governor Ron DiSantis (last name pronounced DeeSantis) has been—with Texas Governor Greg Abbott of Texas—the leading opponent of the use of masks, in schools and elsewhere, in Florida. He prohibited schools from enforcing mask mandates and even sued school districts in his state. Many political pundits believe he has his eye on the White House (as does Greg Abbott) if Donald Trump can not or will not run again.

As a consequence of his anti-common sense health principles, and the rising rate of fatalities in Florida because of Covid-19, some in Florida and elsewhere have begun to call him DeathSantis.

Vice President Pence and his wife tested negative for the virus.

Hospitals in North Carolina begin restricting visitors to their hospitals.

President Donald Trump tweets about the use of Hydroxychloroquine as an effective treatment for the Covid-19 virus. He encouraged the use of that drug so often that late night comics and others suggested he must have stock in the company that manufactures it. The government buys and ships 70,000 doses of Hydroxychloroquine, 10,000 doses of Zithromax and 750,000 doses of chloroquine to New York state, but later studies show Hydroxychloroquine ineffective against Covid-19.

Throughout the year 2020 vaccine deniers gain strongholds among Republican faithful; into 2021 the lowest numbers of those vaccinated are in traditionally Republican southern states.

March 22—Health authorities in New York state recommend health facilities stop testing non-hospitalized patients because of short of PPE (Personal Protective Equipment).

Various states had been in major quarrels regarding purchases of PPE; the states had also been competing with the federal government which had also been buying PPE. These ongoing altercations came to a head early in April, in Washington (see following).

Ohio and West Virginia issue stay-at-home orders; Virginia closes schools for the rest of the academic year.

March 23—Oregon Governor Kate Brown issue a stay-at-home order; New Mexico Governor Michelle Lujan-Grisham closes all non-essential businesses statewide.

Diamond Comic Distributors announces they are no longer accepting any new stock or distributing new comic books until further notice. Because they are the executive distributor, that means all comic books will have to postpone further issues after some titles are released in March.

The WHO describes the pandemic as "accelerating."

The Justice Department creates a price gouging and hoarding task force to investigate cases of scams and other crimes related to the pandemic.

March 24—South Carolina closes all schools from K through high school, through the month of April.

Three U.S. Army hospital units are deployed to New York and Washington State. The Army Corps of Engineers and the National Guard have established four hospitals and four medical centers under construction in New York. FEMA sends 2,000 ventilators to New York.

March 25—Senate Republicans and Democrats strike a deal on a version of the stimulus bill which includes; providing $1,200 to most adults (phased out for people making from $75,000 to $99,000 a year, $600 a week unemployment benefits (approximately $2,400 a month) on top of state unemployment benefits and to last potentially for four months and including freelancers and other workers in the "gig economy" and $500 billion for business and cities.

March 26—The total number of confirmed cases in the United States surpasses that of China with over 85,000, making it the country with the highest number of coronavirus cases in the world.

President Trump announces that the U.S.N.S. ship *Comfort* will be sent to New York City to assist local hospitals; it is due to arrive March 30.

March 27—A survey of 40 economists conducted by the University of Chisago concluded there would be more harm than good in prematurely ending business lockdowns.

President Trump signs the Coronavirus Aid, Relief and Economic Security Act into law.

March 31—Georgia Governor Brian Kemp suspends all in-class instruction for the rest of the academic year.

By the end of March, swabs, test kits, PPE and staff are still in short supply nation-wide.

April 2—Indiana Superintendent of Public Instruction Jenifer McCormick announces that all Indiana schools will close for the rest of the academic year.

Jared Kushner, the best-est son-in-law any Dad could ever have, makes a rare public appearance. He had been tasked with helping supervise the nation's stock of medical supplies and PPE.

As *USA Today* reported,

April 3, 2020, under the headline:
Jared Kushner makes Coronavirus briefing appearance, draws backlash for 'our stockpile' comment

Washington—White House senior advisor Jared Kushner made a rare appearance during Thursday's coronavirus task force briefing, an appearance that drew backlash when he referred to the national stockpile of medical supplies as "our stockpile" and not one belonging to states.

Kushner, the president's son-in-law who doesn't often make public appearances, says he has been serving on the coronavirus task force at the direction of Vice President Mike Pence.

When asked about data showing states' need for equipment, Kushner said "The notion of the federal stockpile is that it's supposed to be our stockpile. It's not supposed to be the states' stockpiles that they can use.

"When you have governors saying that the federal government hasn't given them what they need, I would encourage you to ask them, have you looked within your state to make sure you haven't been able to to find the resources?"

His comments drew immediate blowback. Congressman Ted Lieu, D-Calif, said:

"We are the UNITED STATES of America. The federal stockpile is reserved for all Americans living in our states, not just federal employees. Get it?"

Former White House ethics chief Walter Shaub said:

"It is for the American people … as the federal government's OWN national stockpile website assures us."

The previous website said the stockpile was for use when "local supplies run out," and "when state, local, tribal and territorial responders request federal assistance to support their response efforts."

The website was hastily changed to: "The Strategic National Stockpile's role is to supplement state and local supplies during public health emergencies." It now reads "many states have products stockpiled as well."

President Trump had been defensive about criticism from some governors and mayors who claimed delivery of supplies, including PPE, were very slow.

(And, as an aside, Jared, the best-est son-in-law any Dad could ever have, had been previously tasked to negotiate a permanent peace in the Middle East. You remember how well that went, don't you? Well, don't you?)

April 3—Governor Kay Ivey of Alabama announces a stay-at-home order through April 30.

April 7—President Trump alleges that the WHO mishandled the pandemic. He questioned why the WHO recommended "keeping our borders open to China early on."

April 9—Missouri Governor Mike Parson orders all Missouri schools, public and private, to be closed for in-person instruction for the rest of the year.

April 14—The Associated Press reported that Dr. Anthony Fauci, the government's top infectious disease expert, said Tuesday that the U.SA. does not yet have the critical testing and tracing procedures needed to begin reopening the nation's economy—Fauci said that a May 1 target is "a bit overly optimistic for many areas of the country. Any easing off the social-distancing rules in place in much of the country would have to occur on a rolling basis, not all at once."

As of April 16, there were 639,664 confirmed coronavirus cases in the United States, with 30,985 deaths, a 4.8 percent mortality rate. An estimated 3.24 million tests had been conducted, with indications that 20 percent has tested positive for coronavirus.

April 20—Texas Lieutenant Governor Dan Patrick on the *Tucker Carlson Tonight* show on Fox, said " ... there are more important things than living," as a justification for Texas moving ahead with reopening business despite the coronavirus. Patrick had previously said it would be perfectly appropriate to sacrifice the lives of the elderly for the economy to move ahead. (At that, well more than one skeptic/critic replied: *"Sacrifice the elderly? Your family first."*) Patrick is a staunch Republican conservative; he was previously a right-wing radio talk show host.

April 21—Georgia Governor Brian Kemp announced that many businesses could re-open, including "gyms, hair salons, bowling alleys and tattoo parlors." Opening these close-contact businesses brought outrage from inside and outside of Georgia. Stacey Abrams, the 2018 Democratic candidate for Governor of Georgia, said his position was "dangerously incompetent." As of April 21, Georgia had more than 20,000 confirmed coronavirus cases.

April 30—*Our World in Data* reported that for the U.S., there were 60,966 deaths, 1.04 million confirmed cases and 6.25 million tests completed. The U.S. averaged about 145,200 tests per day between April 1 and April 15, and 199,000 tests per day from April 16 to April 30.

May 6—Maryland Governor Larry Hogan closes all Maryland Public Schools for the rest of the 2019-2020 school year.

May 26—Protests about the George Floyd killing begins in Minneapolis and then throughout the country and, in fact, around the world. *Super-spreader* is used to describe mass gatherings in which participants generally do not wear masks and Covid-19 can be easily spread throughout the crowd.

May 27—Deaths from Covid-19 top 100,000 in the United States.

June 30—*Our World in Data* reports there were 104,140 deaths in the United States, 2.59 million confirmed cases and about 300 million tests completed.

July 27—The United States recorded what was then the largest single-day rise in cases anywhere in the world: 77,638 infections.

July 30—Herman Cain, former Republican candidate for the presidency, had been seen at a Trump rally without a mask, surrounded by others without masks. He died this day from Covid-19, at 74.

August 8—The United States passed five million Covid-19 cases.

August 31—The United States passed six million Covid-19 cases.

September 8—The United States reported less than 25,000 daily cases, for the first time since June.

September 22—The United States surpassed 200,000 deaths from Covid-19.

September 25—The United States passed 7 million cases of Covid-19.

October 2—President Donald Trump and his wife Melania both test positive for Covid-19 after an open air super-spreader event on the White House grounds. He enters Walter Reed National Military Hospital outside Washington, D.C. and is a treated with a myriad of medicines for Covid-19 not available to the public. He leaves and returns to the White House October 5. He later admits in a campaign rally in the south that he got the Covid-19 vaccines; there is audible booing and jeers from his audience.

October 16—The United States passed 8 million Covid-19 cases.

October 30—The United States passes 9 million Covid-19 cases.

November 5—The United States reported a single day record of more than 100,000 new Covid-19 cases.

November 9—The United States passes 10 million Covid-19 cases.

November 15—The United States passes 11 million Covid-19 cases.

November 21—The United States passes 12 million Covid-19 cases, with a record high of more than 200,000 reported in the previous few days.

November 27—The United States passes 13 million Covid-19 cases.

December 2—The United States passes 14 million cases and sets records for the highest number of daily deaths (3,157), new infections (nearly 205K) and hospitalizations (over 100K).

December 8—The United States passes 15 million cases, with one in every 22 Americans having been tested positive since the pandemic began.

December 12—The United States passes 16 million cases.

December 14—The United States surpasses 300,000 deaths, with an average of more than 961 deaths per day since the first known deaths in February. More than 50,000 deaths were reported in the past month.

December 17—The United States passes 17 million cases.

December 21—The United States passes 18 million cases.

December 24—Reports of a new variant of Covid-19 are made public in the United Kingdom.

December 27—The United States a passes 19 million cases, averaging 185,000 cases over the past week.

December 29—A confirmed case of SARS CoV-2 variant from the United Kingdom was reported in Colorado, the first such case in the United States. On December 30 and 31, new variant cases were reported in California and Florida.

By mid-September, 2021, 675,000, or more than approximately 1 in 500 Americans, had died of Covid-19.

In addition to *super-spreader*, two new terms have now entered the public lexicon: *social distancing*, the practice of keeping six feet apart in grocery stores, post offices and any other public buildings, and *long haulers*, those who had Covid-19 and whose symptoms re-appear weeks or months later, on continuing cycles.

Vaccine rates in Republican states, primarily in the South, remain the lowest in the nation, because of political reasons, religious or moral reasons, or for no real reason at all except individual stubbornness.

CNN distributed this headline and article September 20, 2021:

> More people died in Alabama during 2020 than were born as Covid-19 deaths continue to rise
>
> Alabama had more people die last year during the onset of the Covid-19 pandemic than births, a top health official said, as the U.S. continues to grapple with rising coronavirus cases and deaths.
>
> "This past year, 2020, is going to be the first year that we know of in the history our state where we actually have more deaths than births," Alabama Health Officer Dr. Scott Harris said Friday during a news briefing.
>
> "Our state literally shrunk," he said.
>
> Preliminary data shows that 64,714 people died in Alabama in 2020 while there were 57,641 births, Harris said.
>
> In Alabama, where 41 percent of people are fully vaccinated, according to data from the U.S. Centers for Disease Control and Prevention (CDC), officials said more hospital beds have recently become available, though it's not because people are recovering from the virus.
>
> "It is not because these patients are miraculously getting better and going home. It's because they are dying," Dr. Kierstin Kennedy, the chief of hospital medicine at The University of Alabama said.

For some cynics (read: realists) it appears that officials in deep red, southern. Republican states Louisiana, Mississippi, Georgia, Texas and elsewhere, are actively working to kill off their own supporters.

* * *

Lawrence Wright's *The Plague Year: America in the Time of Covid* contains fewer statistics than a timeline of the pandemic, but is no less harrowing.

Early his narrative, he writes:

> During the transition to the Trump administration, the Obama White House handed off a sixty-nine page document called the "Playbook for Early Response to High-Consequence Emerging Infectious Disease Threats and Biological Incidents." A meticulous, step-by-step guide for combatting a "pathogen of pandemic potential," the playbook contains a directory of the government's resources in time of need and is meant to be pulled off the shelf the moment things go haywire.

The playbook outlined the conditions under which, and when, specific government agencies should be employed.

And then what?

Wright says:

> The Trump administration jettisoned the Obama playbook. In 2019, HHS, headed by Alex M. Azar Jr., conducted an exercise called Crimson Contagion. It involved a number of government agencies, including the Pentagon and the NSC; healthcare organizations and major hospitals, such as the Mayo Clinic; public groups with a specific interest in health

care including the American Red Cross; and twelve state governments.

And, he writes:

The Trump administration's own exercise was spookily predictive of what was to come, including how chaotically the government would respond. Federal agencies couldn't tell who was in charge; there was a lack of production capability for personal protective equipment (PPE); ventilators were *in* short supply; and states were frustrated by their attempts to secure enough resources … . The longer the hypothetical contagion went on, the more bollixed the government response became.

And worse, they knew it. The Trump administration had all the best minds in the world in the nation's medical establishment, but … but … but … the results of the Crimson Contagion exercise soon became apparent.

Robert Kadlec, assistant secretary of HHS had led the Crimson Contagion exercise. He later said, "We knew before the movie started it was going to have a bad ending."

Trump proposed cuts to the National Institute of Health; cuts to the CDC and cuts to the Agency for International Development—"the very agencies we needed to fight this outbreak and prevent future ones," Joe Biden said in *USA Today*, January 27, 2020. "And he has treated with utmost contempt institutions that facilitate international cooperation, thus undermining the global efforts that keeps us safe from pandemics and biological attacks."

And Wright says, "The confusion of authority that would characterize every step the Trump administration took in the pandemic was evident from the beginning."

And still, in one of the most infamous statements he ever made, Trump said, "One day, it's like a miracle, it will disappear."

Later, when there were 1,272 cases reported, Wright said, in a remarkable turn of phrase, " ... that was like believing the only stars in the sky were the ones you could see."

And then there was a phrase economics used: the Doom Loop. When there was an economic turndown, and business closed, other factors were triggered. When workers lost their jobs they could not buy food, which weakened supply, which furthered unemployment ... and down and down and down. There was no easily way back up. The Doom Loop.

And when states needed PPE, Trump famously said "try to get it yourselves." He added "we're just the backup. . . . The federal government is not supposed to be out there bu*yi*ng vast amounts of items and then shipping. You know, we're not a shipping clerk."

But that *was exactly* what the states needed: they needed Uncle Sam to *be* the shipping clerk, sending PPE everywhere.

And Donald John Trump refused to wear a mask, believing that was a sign of personal weakness.

And Wright says " ... millions of people followed his example, giving the pandemic access to new communities, infecting new families, endangering health care workers, prolonging unemployment, sabotaging efforts to open the economy and causing untold numbers of people to die."

In short, his vanity caused death.

And Mike Pence, ever the toady, said "Thanks to the leadership of President Trump ... we are winning the fight against the invisible enemy."

Trump sparred with Anthony Fauci time after time; Trump was notoriously—and certainly—anti-science. As Lawrence Wright phrased it:

Covid-19 told us more about these two men than any other individuals in the country. For Fauci, science was a self-correcting compass, always pointed at the truth. For Trump, the truth was Play-doh, and he could twist it to the shape of his desire.

In the first half of 2020, life expectancy fell by a full year, the largest drop since the Second World War, Wright says. At one time there were medical tents set up it New York City's Central Park—the last time that happened was during the Civil War.

Wright's conclusions about Trump are searing:

> By his words and his example, the president became not leader but a saboteur. He ridiculed mask-wearing. He campaigned against taking the pandemic seriously. He subverted his health agencies by installing political operatives who meddled with the science and suppressed the truth. His unmasked political rallies were reckless acts of effrontery. In his Tulsa speech, he frankly admitted that he had told his health officials to "slow the testing down please," so the numbers, the growing constituency of the infected, would be hidden from view. When he finally got the disease, he almost certainly spread it.

Finally, Lawrence Wright wrote:

Covid didn't kill Donald Trump, but it would defeat him.

See also:

The Premonition, Michael Lewis, 2020.
Nightmare Scenario, Yasmeen Abutaleb and Damian Paletta, 2021.

Epilogue

In general parlance, epilogue is taken to mean: *the story is over; it is complete.*

And yet ... and yet ...

There are three quotations that are appropriate here: the first, perhaps by Mr. (or Miss or Mrs.) Anonymous:

> Plagues, like the poor, will always be with us.

The second, cited in the Prologue, by William Faulkner:

> The past is never dead. It's not even past.

The third, and perhaps most chilling, is at the of Lawrence Wright's *The Plague Year: America in the Time of Covid,* and is from Stéphane Bancel, the CEO of Moderna, a firm that manufactures Covid vaccines (emphasis added):

> *We are going to live with this virus, we think, forever.*

References

3   John Howard Griffin in Fensch, *The Man Who Changed His Skin*, pp.14-28.
8   "A bright comet…" and "London at the time …" in Wikipedia entry, "The Great Plague of London."
9   "Outside the city…," Ibid.
286 "An epidemic…" and "This is the classic form…" in Spinney, *Pale Rider*, pp. 95.
288 "from army to army with ease…" Crosby, *American's Forgotten Pandemic*, pp. 25. "Spanish Influenza had …" Ibid. pp. 28.
289 "By May…" Catherine Arnold, *Pandemic 1918*, pp. 57-58.
291 Timeline, History.com, Wikipedia and other sources.
293 "The flu entered Alaska.,.." Spinney 143.
294 "Close to 150…" Ibid. 149.
295 "the flu came," Gina Kolata, Flu: *The Story of the Great Influenza Pandemic of 1918 and the Search for the Virus that Caused It*, … pp. 294.
297 "At three o'clock…" Crosby, pp. 189-190.
298 "A number of people" and "Those who were…" Ibid.. 298.
299 "one of the contributory…" Ibid. 194.
    John Steinbeck's operation in Parini, *John Steinbeck: A Biography*, pp. 19-20.
300 "The psychological damage…" Ibid. pp. 20.
    …" left Hoboken…" Spinney, pp. 47-48.
301 Eleanor Roosevelt, Crosby, pp 115.
    "… in Chicago…" Arnold, 134.
302 Arthur Conan Doyle in Spinney, pp. 237.

John Dos Passos, in Crosby, pp. 315.

Edvard Munch, 'One evening..." in Spinney, pp. 24.

304 "Do you suppose…" Porter, *Pale Horse, Pale Rider*, pp. 207.

306 "Ben's long thin body…," Wolfe, *Look Homeward Angel*, pp. 441.

307 "The only sound…" Ibid., pp. 451.

313 "At seven…" Camus, The Plague, Vintage paperback ed., pp. 288-290.

317 Rock Hudson biography, Wikipedia.

321 Timeline, History,com, Wikipedia and other sources.

337 "…from 1980…" Shilts, pp. xxii.

339 "I am the prettiest one…" Shilts, pp. 21.

340 "It was around this time…" Shilts, pp. 165.

"Whether Gaetan Dugas…" Ibid. pp. 439.

344 "The journey of…" *Journals of the Plague Years*, Spinrad, pp. 141.

"The children of…" Ibid. pp. 142.

345 "Fortunately for me…" Ibid., pp. 146.

348 "We always want…" Jonathan Mann, In Garrett, *The Coming Plague*, pp. xi.

350

356 "Viruses are…" Ibid., pp. 58.

358 Timeline from history.com, Wikipedia and other sources.

380 "During the transition…" Wright, *The Plague Year*, pp. 16.

"The Trump administration…" Ibid., pp. 17.

381 "The Trump administration's…" Ibid, pp. 17.

383 "Covid-19 told us…" Ibid., pp. 191.

"By his words.." Ibid., pp. 243.

"Covid didn't…" Ibid., 245.

385 "We are going to…" Ibid., pp. 269.

Bibliography

Arnold, Catherine. *Pandemic 1918: Eyewitness Accounts from the Greatest Medical Holocaust in Modern History.*
New York: St. Martin's Press, 2018.

Camus, Albert. *The Plague.*
New York: Alfred Knopf, 1948.

Crosby, Alfred W. *America's Forgotten Pandemic: The Influenza of 1918.*
New York: Cambridge University Press, 1989.

Defoe, Daniel. *Journal of the Plague Year.*
London, 1722.

Garrett. Laurie. *The Coming Plague: Newly Emerging Diseases in a World Out of Balance.*
New York: Farrar, Straus and Giroux, 1994.

Kolata, Gina. *Flu: The Story of the Great Influenza Pandemic of 1918 and the Search for the Virus That Caused It.*
New York: Farrar, Straus and Giroux, 1999.

Porter, Katherine Anne. *Pale Horse, Pale Rider.*
New York: Harcourt, Brace, 1936.

Preston, Richard. *The Hot Zone.*
New York: Random House, 1994.

Shuts, Randy. *And the Band Played On: Politics, People and the AIDS Epidemic.*
New York: St. Martin's Press, 1987.
Reprint ed. New York: Penguin Books, 1988.
DVD: HBO Home Video, 1993.

Spinney, Laura. *Pale Rider: The Spanish Flu of 1918 and How It Changed the World.* New York: Public Affairs, 2017.

Spinrad, Norman. *Journals of the Plague Years.* New York: Bantam Books, 1995.

Wolfe, Thomas. *Look Homeward, Angel.* New York: Charles Scribner's Sons, 1929.

Wright, Lawrence. *The Plague Year: America in the Time of Covid.* New York: Knopf, 2021.

About the author …

… a native of Ohio, Thomas Fensch began publishing books in 1970. He is now the author or editor of 41 previous books, including: *Steinbeck and Covici: The Story of a Friendship,* 1979, in print for 40-plus years and long considered a seminal work in John Steinbeck scholarship.

Fensch has also published *The Kennedy-Khrushchev Letters,* 2001; *The Man Who Changed His Skin: The Life and Work of John Howard Griffin, 2011,* the only full biography of Griffin, who wrote the American classic *Black Like Me* and *Orwell in America, 2018.*

Almost all his titles are available on Amazon and elsewhere. Many are in e-book form.

Fensch has a doctorate in print communication from Syracuse University and lives outside Richmond, Virginia, with three 27-inch desktop Apple computers, a working library of 700-plus books and a posse of three dogs, his Senior Literary Advisors, who follow him everywhere and are always ready to fetch yellow pencils, notepads and the like.

www.ingramcontent.com/pod-product-compliance
Lightning Source LLC
Chambersburg PA
CBHW071802080526
44589CB00012B/649